SERIES EDITOR: ALAN SMITH

Modular Maths
for Edexcel

Statistics 1

Second Edition

- **ALAN SMITH**
- **ANTHONY ECCLES, ALAN GRAHAM, NIGEL GREEN, LIAM HENNESSY, ROGER PORKESS**

Hodder & Stoughton

A MEMBER OF THE HODDER HEADLINE GROUP

Acknowledgements

OCR and Edexcel accept no responsibility whatsoever for the accuracy or method of working in the answers given.

All questions acknowledged as MEI are reproduced with kind permission of OCR/MEI.

All questions acknowledged as O & C and Cambridge are reproduced with kind permission of OCR.

Orders: please contact Bookpoint Ltd, 130 Milton Park, Abingdon, Oxon OX14 4SB.
Telephone: (44) 01235 827720, Fax: (44) 01235 400454.
Lines are open from 9.00–5.00, Monday to Saturday, with a 24 hour message answering service.
You can also order through our website www.madaboutbooks.co.uk

British Library Cataloguing in Publication Data
A catalogue record for this title is available from the British Library

ISBN: 978 0 340 88527 7

First published 2000
Second edition published 2004
Impression number   10 9 8 7 6 5 4 3
Year   2010 2009 2008 2007

Copyright in this format © 2000, 2004 Alan Smith

This work includes material adapted from the MEI Structured Mathematics series.

All rights reserved. Apart from any use permitted under UK copyright law, no part of this publication may be reproduced or transmitted in any form or by any means, electronic or mechanical, including photocopy, recording, or any information storage and retrieval system, without permission in writing from the publisher or under licence from the Copyright Licensing Agency Limited. Further details of such licences (for reprographic reproduction) may be obtained from the Copyright Licensing Agency Limited, of Saffron House, 6-10 Kirby Street, London EC1N 8TS.

Papers used in this book are natural, renewable and recyclable products. They are made from wood grown in sustainable forests. The logging and manufacturing processes conform to the environmental regulations of the country of origin.

Cover photo from The Image Bank/Getty Images.
Typeset by Tech-Set Ltd, Gateshead, Tyne & Wear.
Printed in Great Britain for Hodder & Stoughton Educational, a division of Hodder Headline Plc, an Hachette Livre UK Company, 338 Euston Road, London NW1 3BH by Martins the Printers, Berwick upon Tweed.

Edexcel Advanced Mathematics

The Edexcel course is based on units in the four strands of Pure Mathematics, Mechanics, Statistics and Decision Mathematics. The first unit in each of these strands is designated AS, and so is Pure Mathematics: Core 2; all others are A2.

The units may be aggregated as follows:

3 units	AS Mathematics
6 units	A Level Mathematics
9 units	A Level Mathematics + AS Further Mathematics
12 units	A Level Mathematics + A Level Further Mathematics

Core 1 and 2 are compulsory for AS Mathematics, and Core 3 and 4 must also be included in a full A Level award.

Examinations are offered by Edexcel twice a year, in January (most units) and in June (all units). All units are assessed by examination only; there is no longer any coursework in the scheme.

Candidates are not permitted to use electronic calculators in the Core 1 examination. In all other examinations candidates may use any legal calculator of their choice, including graphical calculators.

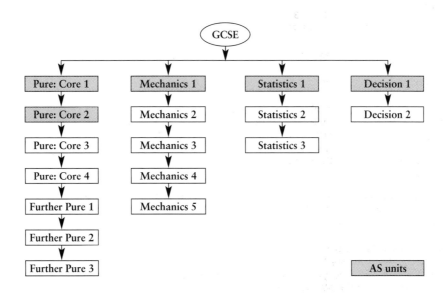

Introduction

This is the first book in a series written to support the Statistics units in the Edexcel Advanced Mathematics scheme. It has been adapted from the successful series written to support the MEI Structured Mathematics scheme, and has been substantially edited and rewritten to provide complete coverage of the new Edexcel Statistics 1 unit.

There are seven chapters in this book. After a short introduction on statistical modelling, there are two chapters on displaying and processing statistical data. You may well be familiar with many of the basic ideas from GCSE, but you should take careful note of how they are refined and implemented in the context of a more advanced mathematics course.

The chapter on probability also builds on ideas you met at GCSE and this is followed by a look at the mathematics of correlation and regression. Discrete random variables are then studied in some detail and the book concludes with an introduction to the normal distribution.

Each chapter contains a number of past examination and examination-style questions.

Throughout the book you will require the use of a calculator, preferably one with a good selection of statistical functions. You should remember that certain calculator restrictions may be enforced by the examination board, but at the time of writing these do not affect any of the Edexcel Statistics units.

I would like to thank the many people who have helped in the preparation and checking of material. Special thanks to Anthony Eccles, Alan Graham, Nigel Green, Liam Hennessy and Roger Porkess, who wrote the original MEI edition, and to Terry Heard for his useful suggestions.

Alan Smith

Contents

1 Mathematical models — 1
Theoretical and experimental probability — 1
Statistical models — 2

2 Representation of data — 5
Categorical or qualitative data — 5
Numerical or quantitative data — 5
Looking at the data — 7
Bar charts and vertical line charts — 14
Pie charts — 17
Histograms — 20
Measures of central tendency and of spread using quartiles — 27
Cumulative frequency curves — 30

3 Summary of data — 42
Measures of central tendency — 42
Frequency distributions — 46
Grouped data — 50
Working with grouped data — 51
Measures of spread — 58
Linear coding — 67

4 Probability — 77
Measuring probability — 77
The complement of an event — 79
Expectation — 80
The probability of either one event or another — 81
Mutually exclusive events — 83
The probability of events from two trials — 87
Conditional probability — 93

5 Correlation and regression — 104

- Describing variables — 105
- Interpreting scatter diagrams — 106
- Line of best fit — 108
- Product moment correlation — 109
- Interpreting correlation — 117
- The least squares regression line — 118
- Linear coding — 126

6 Discrete random variables — 135

- The conditions for a discrete random variable — 136
- Diagrams of probability distributions — 136
- Expectation — 142
- Expectation of a function of X, $E(g[X])$ — 143
- Expectation algebra — 145
- Variance — 148
- The uniform distribution — 156

7 The normal distribution — 163

- Using normal distribution tables — 165
- The normal curve — 168
- Using normal tables in reverse — 172

Answers — 180

Index — 203

Chapter one

Mathematical models

It is a capital mistake to theorize before one has data.
Sir Arthur Conan Doyle

Theoretical and experimental probability

When the outcome of an experiment, or trial, is uncertain we use a *probability* to indicate how likely or unlikely it is. For example, when a fair coin is thrown, the probability of obtaining 'heads' is 0.5, or $\frac{1}{2}$.

The value of a probability lies between 0 and 1. A value of 0 indicates that something cannot happen (impossible) and 1 that it must happen (certain). Probabilities may be written as decimals or fractions, but sometimes one form may be more appropriate than the other.

There are two quite different sorts of probability.

Theoretical probability

To calculate a theoretical probability you have to decide how many equally likely outcomes an experiment might produce. Some of those outcomes are favourable to a particular event, say A. Then

$$\text{Probability } (A \text{ occurs}) = \frac{\text{number of outcomes favourable to the event } A}{\text{total number of all possible outcomes}}$$

EXAMPLE 1.1

A fair die is thrown. Find the probability that the score is greater than 2.

Solution There are 6 equally likely outcomes, namely 1, 2, 3, 4, 5, 6. Of these there are four which are favourable to the event 'the score is greater than 2', namely 3, 4, 5 and 6. Therefore

$$\text{Probability (score is greater than 2)} = \tfrac{4}{6}$$
$$= \tfrac{2}{3}$$

EXPERIMENTAL PROBABILITY

To calculate an experimental probability you have to carry out an experiment first, recording the number of times the outcome was favourable to a particular event A. Then

$$\text{Probability } (A \text{ occurs}) = \frac{\text{number of times that an outcome favourable to } A \text{ occurred}}{\text{total number of trials}}$$

EXAMPLE 1.2

Dean has a die which he suspects might not be fair. He throws it 600 times, with the following results:

Score, X	1	2	3	4	5	6
Number of times that X occurred	108	94	91	103	95	109

Find the probability that the score is greater than 4.

Solution The event 'the score is greater than 4' occurred $95 + 109 = 204$ times.

$$\text{Probability (score is greater than 4)} = \tfrac{204}{600}$$
$$= 0.34$$

STATISTICAL MODELS

The theoretical probability solution to Example 1.1 assigns an equal probability to each of the possible outcomes. This gives rise to a *discrete uniform distribution* which you will study in further detail in Chapter 6. It is one of the standard *statistical models* used in AS and A level work.

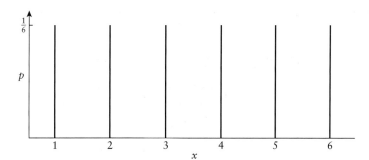

FIGURE 1.1

Another widely encountered model is the *normal distribution*. This is used to describe continuous variables, such as heights or weights, and is only valid provided they are distributed according to a certain bell-shaped mathematical rule:

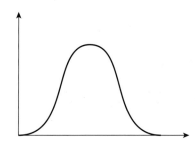

FIGURE 1.2

Note that all normal distributions must be *symmetrical*. The normal distribution is thus a poor model in situations where the distribution is strongly *skewed*, that is possesses an extended tail in one direction:

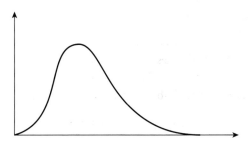

FIGURE 1.3

EXERCISE 1A

Say whether you think each of these discrete variables may be modelled by a discrete uniform distribution. Give a brief reason for each answer.

1 The number of heads obtained when a coin is tossed once.

2 The number of heads obtained when two coins are tossed together.

3 The total score obtained when two dice are thrown.

4 The score showing on a fair four-sided spinner labelled 1, 2, 3, 4.

5 The number of university students born in each of the months January, February, etc.

6 The number of marks achieved by students, out of 10, in a short arithmetic test.

7 The number of matches in a box which is labelled 'Average contents 50'.

8 The number of goals scored by a soccer team last season.

9 The number of brothers and sisters that each member of Class 5B has.

10 The number of minutes of 'injury time' added on at the end of a World Cup soccer match.

Say whether you think each of these continuous variables may be modelled by a normal distribution. Give a brief reason for each answer.

11 The masses of cucumbers grown on a farm.

12 The ages at which people in the UK got married last year.

13 The times taken by pupils to run a local school cross-country race.

14 The lengths of baby lizards at age 7 days.

15 The weights of Year 7 pupils at a local school.

16 The exact amount of cola in a '330 ml' can.

17 The lengths, in seconds, of all the CD tracks in my collection.

18 The ages at which people retired from work in the UK last year.

19 The time it takes me to walk to work each morning.

20 The mass of tea in a particular brand of tea bag.

You will meet these and other statistical models as you work through the various statistics units in the course.

Whenever a model is encountered you should ask yourself how appropriate the model is in that specific situation, and what simplifying assumptions might be implied in your choice of model.

For example, in assigning a probability of 0.5 to each of 'heads' and 'tails' when a coin is tossed we are assuming that it cannot land on end!

Chapter two

REPRESENTATION OF DATA

A picture is worth a thousand words.

Anon

CATEGORICAL OR QUALITATIVE DATA

Some data come to you in classes or categories. Such data, like these for the members of the Select Committee for Education and Employment, are called *categorical*, or *qualitative*.

L L L C L   L L L C L   L D C L L C   L LD

C = Conservative; L = Labour; LD = Liberal Democrat
Members are listed alphabetically.
(Source: *www.parliament.uk* August 1999)

Most of the data you encounter, however, will be *numerical* data (also called *quantitative* data).

NUMERICAL OR QUANTITATIVE DATA

Variable

The score you get when you throw an ordinary die is one of the values 1, 2, 3, 4, 5 or 6. Rather than repeatedly using the phrase 'the score you get when you throw an ordinary die', statisticians find it convenient to use a capital letter, X, say. They let X stand for 'the score you get when you throw an ordinary die' and because this varies, X is referred to as a *variable*.

Similarly, if you are collecting data and this involves, for example, noting the temperature in classrooms at noon, then you could let T stand for 'the temperature in a classroom at noon'. So T is another example of a variable.

Random variable

If a variable has an associated *probability*, such as when throwing an ordinary die the probability of getting a 6 is $\frac{1}{6}$, then the variable is referred to as a *random variable*.

It is possible to record most data as values of a *variable*, for example, people's heights or weights, or the score on a die. In such cases, the classification is done on the basis of a number, rather than a description, and the data are *numerical* or *quantitative*. Numerical or quantitative data are either *discrete* or *continuous*.

DISCRETE AND CONTINUOUS VARIABLES

The scores on a die, 1, 2, 3, 4, 5 and 6, the number of goals a football team scores, 0, 1, 2, 3, ... and British shoe sizes 1, $1\frac{1}{2}$, 2, $2\frac{1}{2}$, ... are all examples of *discrete variables*. What they have in common is that all possible values can be listed.

Distance, mass, temperature and speed are all examples of *continuous variables*. Continuous variables, if measured accurately enough, can take any appropriate value. You cannot list all possible values.

Age is rather a special case. It is nearly always given rounded down (i.e. truncated). Although your age changes continuously every moment of your life, you actually state it in steps of one year, in completed years, and not to the nearest whole year. So a man who is a few days short of his 20th birthday will still say he is 19.

In practice, data for a continuous variable are always given in a rounded form.

- A person's height, h, is given as 168 cm, measured to the nearest centimetre; $167.5 \leqslant h < 168.5$
- A temperature, T, is given as 21.8 °C, measured to the nearest tenth of a degree; $21.75 \leqslant T < 21.85$
- The depth of an ocean, d, is given as 9200 m, measured to the nearest 100 m; $9150 \leqslant d < 9250$

Notice the rounding convention here: if a figure is on the borderline it is rounded up. There are other rounding conventions.

Raw data

A newspaper reporter wanted information about local road accidents involving cyclists.

His assistant collected the information in this form:

Name	Age	Distance from home	Cause	Injuries	Treatment
John Smith	45	3 km	Skid	Concussion	Hosp. outpatient
Debbie Lane	6	75 km	Hit kerb	Broken arm	Hosp. outpatient
Arvinder Sethi	12	1200 m	Lorry	Multiple fractures	Hosp. 3 weeks
Marion Wren	8	300 m	Hit each other	Bruising	Hosp. outpatient
David Huker	8	50 m		Concussion	Hosp. overnight

There were 92 accidents listed in the reporter's table.

Ages of cyclists (from survey)

66	6	62	19	20	15	21	8	21	63	44	10	44	34	18
35	26	61	13	61	28	21	7	10	52	13	52	20	17	26
64	11	39	22	9	13	9	17	64	32	8	9	31	19	22
37	18	138	16	67	45	10	55	14	66	67	14	62	28	36
9	23	12	9	37	7	36	9	88	46	12	59	61	22	49
18	20	11	25	7	42	29	6	60	60	16	50	16	34	14
18	15													

This information is described as *raw data*, which means that no attempt has yet been made to organise it in order to look for any patterns.

LOOKING AT THE DATA

At the moment the arrangement of the ages of the 92 cyclists tells you very little at all. Clearly these data must be organised so as to reveal the underlying shape, the *distribution*. The figures need to be ranked according to size and preferably grouped as well. The reporter had asked an assistant to collect the information and this was the order in which she presented it.

TALLY

Tallying is a quick, straightforward way of grouping data into suitable intervals. You have probably met it already.

Stated age (years)	Tally	Frequency
0–9	ЖТ ЖТ III	13
10–19	ЖТ ЖТ ЖТ ЖТ I	26
20–29	ЖТ ЖТ ЖТ I	16
30–39	ЖТ ЖТ	10
40–49	ЖТ I	6
50–59	ЖТ	5
60–69	ЖТ ЖТ IIII	14
70–79		0
80–89	I	1
⋮		
130–139	I	1
	TOTAL	92

EXTREME VALUES

A tally immediately shows up any extreme values, that is values which are far away from the rest. In this case there are two extreme values, usually referred to as *outliers*: 88 and 138. Before doing anything else you must investigate these.

In this case the 88 is genuine, the age of Millie Smith, who is a familiar sight cycling to the shops.

The 138 needless to say is not genuine. It was the written response of a man who was insulted at being asked his age. Since no other information about him is available, this figure is best ignored and the sample size reduced from 92 to 91. You should always try to understand an outlier before deciding to ignore it; it may be giving you important information.

Practical statisticians are frequently faced with the problem of outlying observations, *observations that depart in some way from the general pattern of a data set. What they, and you, have to decide is whether any such observations belong to the data set or not. In the above example the data value 88 is a genuine member of the data set and is retained. The data value 138 is not a member of the data set and is therefore rejected.*

Describing the shape of a distribution

An obvious benefit of using a tally is that it shows the overall shape of the distribution.

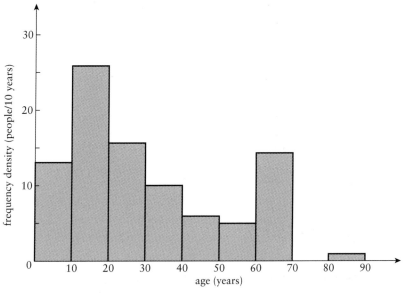

FIGURE 2.1 *Histogram to show the ages of people involved in cycling accidents*

You can now see that a large proportion (more than a quarter) of the sample are in the 10 to 19 year age range. This is the *modal* group as it is the one with the most members. The single value with the most members is called the *mode*, in this case age 9.

You will also see that there is a second peak among those in their sixties; so this distribution is called *bimodal*, even though the frequency in the interval 20–29 is greater than the frequency in the interval 60–69.

Different types of distribution are described in terms of the position of their modes or modal groups, see figure 2.2.

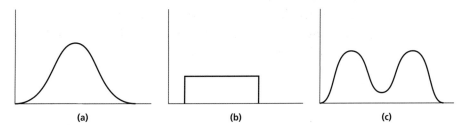

FIGURE 2.2 *Distribution shapes:*
(a) *unimodal and symmetrical* **(b)** *uniform (no mode but symmetrical)* **(c)** *bimodal*

When the mode is off to one side the distribution is said to be *skewed*. If the mode is to the left with a long tail to the right the distribution has positive or right skewness; if the long tail is to the left the distribution has negative or left skewness. These two cases are shown in figure 2.3.

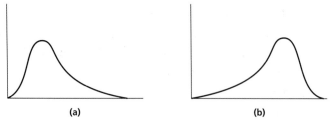

FIGURE 2.3 Skewness: **(a)** *positive* **(b)** *negative*

STEM AND LEAF DIAGRAMS OR STEMPLOTS

The quick and easy view of the distribution from the tally has been achieved at the cost of losing information. You can no longer see the original figures which went into the various groups and so cannot, for example, tell from looking at the tally whether Millie Smith is 80, 81, 82, or any age up to 89. This problem of the loss of information can be solved by using a *stem and leaf diagram* (or *stemplot*).

This is a quick way of grouping the data so that you can see their distribution and still have access to the original figures. The one below shows the ages of the 91 cyclists surveyed.

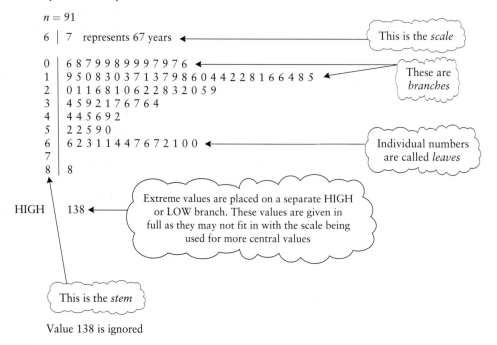

Value 138 is ignored

FIGURE 2.4 *Stem and leaf diagram showing the ages of a sample of 91 cyclists (unsorted)*

LOOKING AT THE DATA

The column of figures on the left (going from 0 to 8) corresponds to the tens digits of the ages. This is called the *stem* and in this example it has 9 branches. On each branch on the stem are the *leaves* and these represent the units digits of the data values.

As you can see above, the leaves for a particular branch have been placed in the order in which the numbers appeared in the original raw data. This is fine for showing the general shape of the distribution, but it is usually worthwhile sorting the leaves, as shown in figure 2.5.

$n = 91$

6 | 7   represents 67 years

```
0 | 6 6 7 7 7 8 8 9 9 9 9 9 9
1 | 0 0 0 1 1 2 2 3 3 3 3 4 4 4 4 5 5 6 6 6 6 7 7 8 8 8 8 9 9
2 | 0 0 0 1 1 1 2 2 2 2 3 5 6 6 8 8 9
3 | 1 2 4 4 5 6 6 7 7 9
4 | 2 4 4 5 6 9
5 | 0 2 2 5 9
6 | 0 0 1 1 1 2 2 3 4 4 6 6 7 7
7 |
8 | 8
```

Note that the value 138 is left out as it has been identified as not belonging to this set of data.

FIGURE 2.5 *Stem and leaf diagram showing the ages of a sample of 91 cyclists (sorted)*

The stem and leaf diagram gives you a lot of information at a glance:

- The youngest cyclist is 6 and the oldest is 88 years of age
- More people are in the 10–19 year age range than in any other 10 year age range
- There are three 61 year olds
- The modal age (i.e. the age with the most people) is 9.

If the values on the basic stem and leaf diagram are too cramped, that is, if there are so many leaves on a line that the diagram is not clear, you may *stretch* it.
To do this you put values 0, 1, 2, 3, 4 on one line and 5, 6, 7, 8, 9 on another. Doing this to the example produces the diagram shown in figure 2.6.

When stretched, this stem and leaf diagram reveals the skewed nature of the distribution.

$n = 91$

6 | 7 represents 67

```
0  |
0  | 6 6 7 7 7 8 8 9 9 9 9 9 9
1* | 0 0 0 1 1 2 2 3 3 3 4 4 4
1  | 5 5 6 6 6 7 7 8 8 8 8 9 9
2* | 0 0 0 1 1 1 2 2 2 3
2  | 5 6 6 8 8 9
3* | 1 2 4 4
3  | 5 6 6 7 7 9
4* | 2 4 4
4  | 5 6 9
5* | 0 2 2
5  | 5 9
6* | 0 0 1 1 1 2 2 3 4 4
6  | 6 6 7 7
7* |
7  |
8* |
8  | 8
```

You must include all the branches, even those with no leaves

FIGURE 2.6 *Stem and leaf diagram showing the ages of a sample of 91 cyclists (sorted)*

Stem and leaf diagrams are particularly useful for comparing data sets.
With two data sets a back-to-back stem and leaf diagram can be used, as shown in figure 2.7.

represents 590   9 | 5 | 2   represents 520

```
           9 | 5 | 1 7
           2 | 6 | 0 2 3 5 8
       5 3 0 | 7 | 1 2 5 6 6 7
     9 7 5 1 1 | 8 | 3 5
       8 6 2 1 | 9 | 2
```

Note that the numbers on the left of the stem still have the smallest number next to the stem

FIGURE 2.7

Exercise 2A

1 Write down the numbers which are represented by this stem and leaf diagram.

$n = 15$

```
32 | 1      represents 3.21 cm
32 | 7
33 | 2 6
34 | 3 5 9
35 | 0 2 6 6 8
36 | 1 1 4
37 | 2
```

2 Write down the numbers which are represented by this stem and leaf diagram.

$n = 19$

```
 8 | 9     represents 0.089 mm
 8 | 3 6 7
 9 | 0 1 4 8
10 | 2 3 5 8 9 9
11 | 0 1 4
12 | 3 5
13 | 1
```

3 Show the following numbers on a stem and leaf diagram with six branches, remembering to include the appropriate scale.

0.212   0.223   0.226   0.230   0.233   0.237   0.241
0.242   0.248   0.253   0.253   0.259   0.262

4 Show the following numbers on a stem and leaf diagram with five branches, remembering to include the appropriate scale.

81.07   82.00   78.01   80.08   82.05
81.09   79.04   81.03   79.06   80.04

5 Write down the numbers which are represented by this stem and leaf diagram.

$n = 21$

```
34 | 5     represents 3.45 m
LOW   0.013, 0.089, 1.79
34 | 3
35 | 1 7 9
36 | 0 4 6 8
37 | 1 1 3 8 9
38 | 0 5
39 | 4
HIGH   7.42, 10.87
```

6 Forty motorists entered for a driving competition. The organisers were anxious to know if the contestants had enjoyed the event and also to know their ages, so that they could plan and promote future events effectively. They therefore asked entrants to fill in a form on which they commented on the various tests and gave their ages.

The information was copied from the forms and the ages listed as:

    28 52 44 28 38   46 62 59 37 60
    19 55 34 35 66   37 22 26 45 5
    61 38 26 29 63   38 29 36 45 33
    37 41 39 81 35   35 32 36 39 33

(a) Plot these data as an unsorted stem and leaf diagram.
(b) Identify any outliers and comment on them.

7 The unsorted stem and leaf diagram below gives the ages of males whose marriages were reported in a local newspaper one week.

$n = 42$

1 | 9   represents 19

```
0 |
1 | 9 6 9 8
2 | 5 6 8 9 1 1 0 3 6 8 4 1 2 7
3 | 0 0 5 2 3 9 1 2 0
4 | 8 4 7 9 6 5 3 3 5 6
5 | 2 2 1 7
6 |
7 |
8 | 3
```

(a) Identify and comment on any outliers.
(b) What was the age of the oldest person whose marriage is included?
(c) Redraw the stem and leaf diagram with the leaves sorted.
(d) Stretch the stem and leaf diagram by using steps of five years between the levels rather than ten.
(e) Describe and comment on the distribution.

Bar charts and vertical line charts

It is best to use bar charts to illustrate categorical data and vertical line charts to illustrate discrete data, although people on occasion use them the other way round. The height of each bar or line represents the frequency.

If bars are used there should be gaps between the bars. The widths and areas of the bars have no significance, but all the bars should be the same width to avoid distorting the picture of the data.

The political parties of the members of the Select Committee for Education and Employment, August 1999 are represented in the bar chart shown in figure 2.8. It is immediately obvious that the Labour party has a majority membership of this committee.

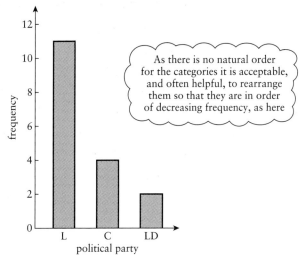

As there is no natural order for the categories it is acceptable, and often helpful, to rearrange them so that they are in order of decreasing frequency, as here

FIGURE 2.8   *A bar chart illustrating categorical data*

Rachel plays cricket for the local ladies' cricket club. During her first season, in which she batted and bowled for the team, she summarised her batting record in the following diagram.

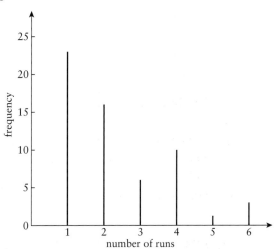

FIGURE 2.9   *A vertical line chart illustrating discrete data*

Even though a bar chart can be used in this example, a line chart is preferable as it shows quite clearly that the scores can only take integer values.

There are many different ways of drawing bar charts. The bars can be horizontal or vertical. The bars can also be subdivided. A compound bar chart is shown in figure 2.10. Often there is no single right way of displaying the information; what is most important is that it should be easy to follow and not misleading.

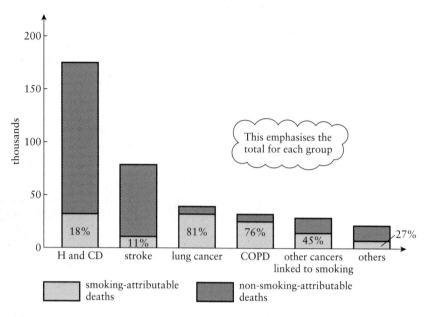

FIGURE 2.10　*Compound bar chart showing smoking- and non-smoking-attributable deaths* (Source: *The Smoking Epidemic*, HEA, 1991)

Figure 2.11 shows a multiple bar chart comparing the level of sales of three products of a company over a period of four years. Note that there is a gap between the information for each year.

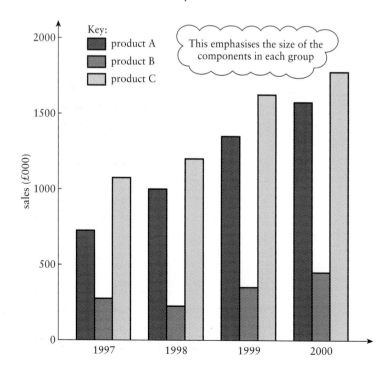

FIGURE 2.11

EXERCISE 2B

PIE CHARTS

A pie chart can be used to illustrate categorical (or qualitative) data or it can be used to illustrate discrete or grouped continuous data. Pie charts are used to show the size of constituent parts relative to the whole.

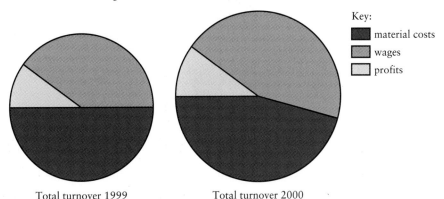

Key:
■ material costs
▨ wages
□ profits

Total turnover 1999   Total turnover 2000

FIGURE 2.12

An electronics firm increased their turnover from £1 800 000 in 1999 to £2 400 000 in 2000. The pie charts show the division of the money between wages, material costs and profits.

The increase in total turnover from 1999 to 2000 is reflected in the larger area of the second pie chart.

The values of the data are proportional to the areas of the pie charts and not proportional to the radii.

For example: Area of chart 2 = $\frac{4}{3}$ × area of chart 1

EXERCISE 2B

1 (i) State whether the data described below are categorical or numerical, and, if numerical, whether discrete or continuous.
  (ii) State what you think would be the most appropriate method of displaying the data.
  (a) The number of coins in shoppers' purses
  (b) The colour of the eyes of a sample of people
  (c) The masses of a sample of eggs
  (d) The medals (gold, silver and bronze) won by a team at the Olympic games
  (e) The sizes of a sample of eggs (size 1, size 2, etc.)
  (f) The times of the runners in a 100-metre race
  (g) The numbers on the shirts of a sample of rugby players
  (h) The scores of the competitors in an ice-skating competition
  (i) Estimates of the length of a needle
  (j) The single letter on the registration plate of a sample of cars.

2 The two pie charts are drawn to scale. They represent the income of two North Atlantic islands one year. The total income for Seanna was £72 000. Calculate the earnings for both islands from each product.

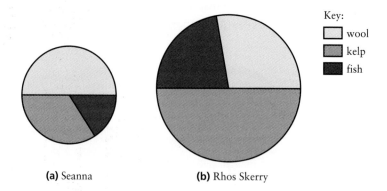

(a) Seanna (b) Rhos Skerry

Key:
- wool
- kelp
- fish

3 A developing country's education budget is divided between the primary, secondary and tertiary sectors. The figures below give the amount of money spent on each sector. They are in millions of dollars and have been adjusted to allow for inflation.

	Primary	Secondary	Tertiary
1960	4.1	1.1	0.9
1975	12.2	8.4	6.0
1990	18.0	20.8	16.9

(a) Draw compound bar charts to illustrate this information.
(b) Draw, to scale, three pie charts to illustrate the information.
(c) Comment briefly on what the figures tell you about the country's education programme.
(d) Which do you consider to be the better method of display and why?

4 The compound bar chart shows the production of three cash crops from a region of an African country in the years 1990 and 2000.

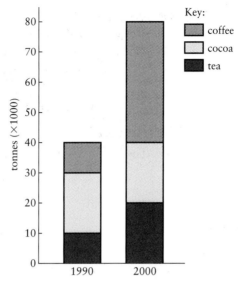

Key:
- coffee
- cocoa
- tea

(a) Draw, to scale, two pie charts to illustrate the same information.
(b) Comment on the changes in crop production over the ten years.
(c) Which do you consider to be the better method of display and why?

5 The following table shows the breakdown of information for *Workright Co. Ltd* over a four-year period.

	1997 (£m)	1998 (£m)	1999 (£m)	2000 (£m)
Turnover	4.0	4.5	5.8	6.0
Wages	1.3	1.9	2.2	2.5
Production costs	0.4	0.5	0.6	0.8
Material costs	1.0	1.3	1.6	1.8
Taxation	0.2	0.1	0.3	0.0
Other costs	0.5	0.5	0.7	0.8
Profit	0.6	0.2	0.4	0.1

(a) Illustrate the above data using an appropriate diagram.
(b) Comment on the company's performance.

6 Data for *Camford Electronics* for 1999 and 2000 are shown in the table below.

	Turnover	Material costs	Wages	Profits
1999	£1 800 000	£900 000	£720 000	£180 000
2000	£2 400 000	£1 100 000	£1 100 000	£200 000

Draw an appropriate bar chart to compare the data for 1999 and 2000.

Histograms

Histograms are used to illustrate continuous data. The columns in a histogram may have different widths and it is the area of each column which is proportional to the frequency and not the height. Unlike bar charts, there are no gaps between the columns because where one class ends the next begins.

Continuous data with equal class widths

A sample of 60 components is taken from a production line and their diameters, d mm, recorded. The resulting data are summarised in the following frequency table.

Length (mm)	Frequency
$25 \leqslant d < 30$	1
$30 \leqslant d < 35$	3
$35 \leqslant d < 40$	7
$40 \leqslant d < 45$	15
$45 \leqslant d < 50$	17
$50 \leqslant d < 55$	10
$55 \leqslant d < 60$	5
$60 \leqslant d < 65$	2

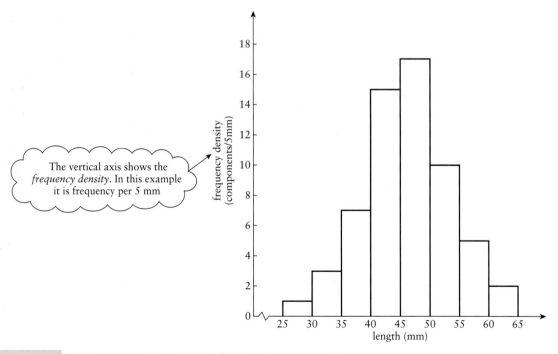

The vertical axis shows the *frequency density*. In this example it is frequency per 5 mm

Figure 2.13   *Histogram to show the distribution of component diameters*

The class boundaries are 25, 30, 35, 40, 45, 50, 55, 60 and 65. The width of each class is 5.

The area of each column is proportional to the class frequency. In this example the class widths are equal so the height of each column is also proportional to the class frequency.

The column representing $45 \leq d < 50$ is the highest and this tells you that this is the modal class, that is, the class with highest frequency per 5 mm.

LABELLING THE FREQUENCY AXIS

The vertical axis tells you the frequency *density*. Figure 2.15 looks the same as 2.14 but it is not a histogram. This type of diagram is, however, often incorrectly referred to as a histogram. It is more correctly called a frequency chart.
A histogram shows the frequency density on the vertical axis.

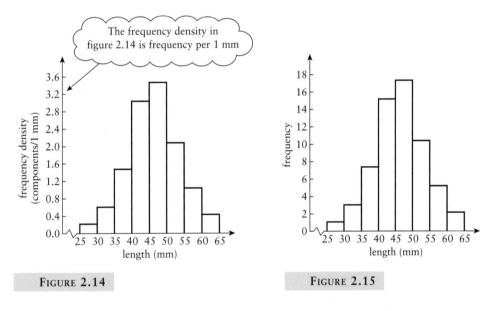

FIGURE 2.14   FIGURE 2.15

The units which you choose for the frequency density are particularly important when the class widths are unequal, as you will see in the next example.

CONTINUOUS DATA WITH UNEQUAL CLASS WIDTHS

The heights of 80 broad bean plants were measured, correct to the nearest centimetre, ten weeks after planting. The data are summarised in the frequency table.

Height (cm)	Frequency	Class width (cm)	Frequency per 2 cm
$7.5 \leqslant x < 11.5$	1	4	$\frac{1}{2}$
$11.5 \leqslant x < 13.5$	3	2	3
$13.5 \leqslant x < 15.5$	7	2	7
$15.5 \leqslant x < 17.5$	11	2	11
$17.5 \leqslant x < 19.5$	19	2	19
$19.5 \leqslant x < 21.5$	14	2	14
$21.5 \leqslant x < 23.5$	13	2	13
$23.5 \leqslant x < 25.5$	9	2	9
$25.5 \leqslant x < 28.5$	3	3	2

Most of the classes are 2 cm wide so it is convenient to take 2 cm as the *standard width*.

The first class is twice the standard width; consequently the height of this column on the histogram is half the given frequency. The last class is $\frac{3}{2}$ times the standard width so the height of the column is $\frac{2}{3}$ of the given frequency. The area of each column is proportional to the class frequency.

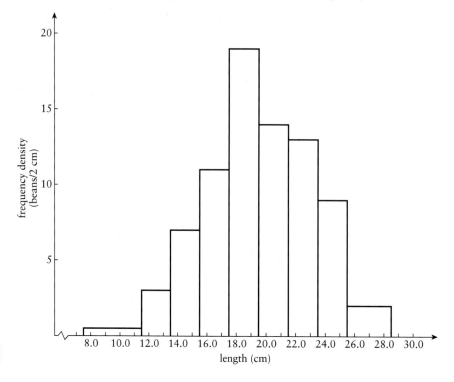

FIGURE 2.16

Discrete data

 Histograms are occasionally used for grouped discrete data. However, you should always first consider the alternatives.

A test was given to 100 students. The maximum mark was 70. The raw data are shown below.

```
10  18  68  67  25  62  49  11  12   8
 9  46  53  57  30  63  34  21  68  31
20  16  29  13  31  56   9  34  45  55
35  40  45  48  54  50  34  32  47  60
70  52  21  25  53  41  29  63  43  50
40  48  45  38  51  25  52  55  47  46
46  50   8  25  56  18  20  36  36   9
38  39  53  45  42  42  61  55  30  38
62  47  58  54  59  25  24  53  42  61
18  30  32  45  49  28  31  27  54  38
```

Illustrating this data using a vertical line graph results in the following:

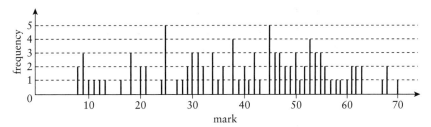

FIGURE 2.17

This diagram fails to give a clear picture of the overall distribution of marks. In this case you could consider a bar chart or, as the individual marks are known, a stem and leaf diagram, as follows.

$n = 100$

2 | 5   represents 25 marks

```
0 | 8 8 9 9 9
1 | 0 1 2 3 6 8 8 8
2 | 0 0 1 1 4 5 5 5 5 5 7 8 9 9
3 | 0 0 0 1 1 1 2 2 4 4 4 5 6 6 8 8 8 9
4 | 0 0 1 2 2 2 3 5 5 5 5 5 6 6 6 7 7 7 8 8 9 9
5 | 0 0 0 1 2 2 3 3 3 3 4 4 4 5 5 5 6 6 7 8 9
6 | 0 1 1 2 2 3 3 7 8 8
7 | 0
```

FIGURE 2.18

If the data have been grouped and the original data have been lost, or are otherwise unknown, then a histogram may be considered. A grouped frequency table and histogram illustrating the marks are shown below.

Marks, x	Frequency, f
0–9	5
10–19	8
20–29	14
30–39	19
40–49	22
50–59	21
60–70	11

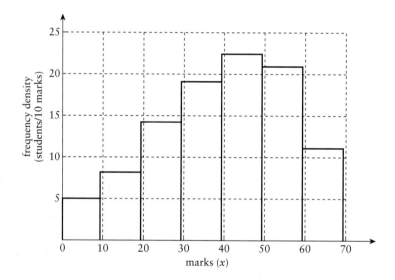

FIGURE 2.19

Note The class boundary 10–19 becomes $9.5 \leqslant x < 19.5$ for the purpose of drawing the histogram. You must give careful consideration to class boundaries, particularly if you are using rounded data.

Grouped discrete data are illustrated well by a histogram if the distribution is particularly skewed as is the case in the next example.

The first 50 positive integers squared are:

```
   1    4    9   16   25   36   49   64
  81  100  121  144  169  196  225  256
 289  324  361  400  441  484  529  576
 625  676  729  784  841  900  961 1024
1089 1156 1225 1296 1369 1444 1521 1600
1681 1764 1849 1936 2025 2116 2209 2304
2401 2500
```

EXERCISE 2C

Number, n	Frequency, f
$0 < n \leq 250$	15
$250 < n \leq 500$	7
$500 < n \leq 750$	5
$750 < n \leq 1000$	4
$1000 < n \leq 1250$	4
$1250 < n \leq 1500$	3
$1500 < n \leq 1750$	3
$1750 < n \leq 2000$	3
$2000 < n \leq 2250$	3
$2250 < n \leq 2500$	3

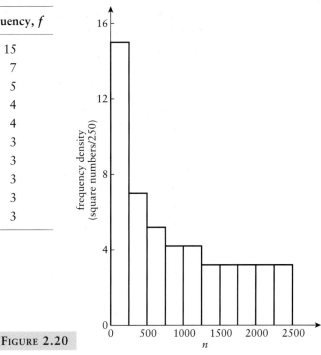

FIGURE 2.20

The main points to remember when drawing a histogram are:

- Histograms are usually used for illustrating continuous data. For discrete data it is better to draw a stem and leaf diagram, line graph or bar chart.
- Since the data are continuous, or treated as if they were continuous, adjacent columns of the histogram should touch (unlike a bar chart where the columns should be drawn with gaps between them).
- It is the areas and not the heights of the columns that are proportional to the frequency of each class.
- The vertical axis should be marked with the appropriate frequency density (*frequency per 5 mm* for example), rather than frequency.

EXERCISE 2C

1 A number of trees in two woods were measured. Their diameters, correct to the nearest centimetre, are summarised in the table below.

Diameter (cm)	1–10	11–15	16–20	21–30	31–50	Total
Akeley Wood	10	5	3	11	1	30
Shaw's Wood	6	8	20	5	1	40

(Trees less than $\frac{1}{2}$ cm in diameter are not included.)
(a) Write down the actual class boundaries.
(b) Draw two separate histograms to illustrate this information.
(c) State the modal class for each wood.
(d) Describe the main features of the distributions for the two woods.

2   Listed below are the prime numbers, p, from 1 up to 1000. (1 itself is not usually defined as a prime.)

Primes up to 1000

2	3	5	7	11	13	17	19	23	29	31	37	41	43
47	53	59	61	67	71	73	79	83	89	97	101	103	107
109	113	127	131	137	139	149	151	157	163	167	173	179	181
191	193	197	199	211	223	227	229	233	239	241	251	257	263
269	271	277	281	283	293	307	311	313	317	331	337	347	349
353	359	367	373	379	383	389	397	401	409	419	421	431	433
439	443	449	457	461	463	467	479	487	491	499	503	509	521
523	541	547	557	563	569	571	577	587	593	599	601	607	613
617	619	631	641	643	647	653	659	661	673	677	683	691	701
709	719	727	733	739	743	751	757	761	769	773	787	797	809
811	821	823	827	829	839	853	857	859	863	877	881	883	887
907	911	919	929	937	941	947	953	967	971	977	983	991	997

(a) Draw a histogram to illustrate these data with the following class intervals:
$1 \leqslant p < 20 \quad 20 \leqslant p < 50 \quad 50 \leqslant p < 100 \quad 100 \leqslant p < 200$
$200 \leqslant p < 300 \quad 300 \leqslant p < 500$ and $500 \leqslant p < 1000$.

(b) Comment on the shape of the distribution.

3   A case containing 270 oranges was opened and each orange was weighed to the nearest gram. The masses were found to be distributed as in the following table:

Mass (grams)	Number of oranges
60–99	20
100–119	60
120–139	80
140–159	50
160–219	60

Draw a histogram to illustrate the data.

4 In an agricultural experiment, 320 plants were grown on a plot, and the lengths of the stems were measured to the nearest centimetre ten weeks after planting. The lengths were found to be distributed as in the following table:

Length (cm)	Number of plants
20–31	30
32–37	80
38–43	90
44–49	60
50–67	60

Draw a histogram to illustrate the data.

5 The lengths of time of sixty songs recorded by a certain group of singers are summarised in the table below:

Song length in seconds (x)	Number of songs
$0 < x < 120$	1
$120 \leqslant x < 180$	9
$180 \leqslant x < 240$	15
$240 \leqslant x < 300$	17
$300 \leqslant x < 360$	13
$360 \leqslant x \leqslant 600$	5

Display the data on a histogram.

MEASURES OF CENTRAL TENDENCY AND OF SPREAD USING QUARTILES

You should already know how to find the median of a set of discrete data. As a reminder, the median is the value of the middle item when all the data items have been ranked in order.

The median is the value of the $\frac{n+1}{2}$ th item and is half-way through the data set. The values one-quarter of the way through the data set and three-quarters of the way through the data set are called the *lower quartile* and the *upper quartile* respectively. The lower quartile, median and upper quartile are usually denoted using Q_1, Q_2 and Q_3.

Quartiles are used mainly with large data sets and their values found by looking at the $\frac{1}{4}$, $\frac{1}{2}$ and $\frac{3}{4}$ points. So, for a data set of 1000, you would take Q_1 to be the value of the 250th data item, Q_2 to be the value of the 500th data item and Q_3 to be the value of the 750th data item.

It's better to avoid using quartiles with small data sets or samples since the value of any of them is heavily dependent on one or two members of the set. If you cannot avoid working with a small sample then the median is the value of the $\frac{n+1}{2}$th item. If $\frac{n+1}{2}$ is a whole number value, m, you use $\frac{m+1}{2}$ to find the position of the lower quartile. So, for a data set of nine items Q_2 is the value of the fifth item ($\frac{9+1}{2}$). Q_1 is the value of the third item ($\frac{5+1}{2}$) from the bottom of the data set and Q_3 is the value of the third item from the top of the data set.

If $\frac{n+1}{2}$ is a half value, ignore the half to find Q_1 and Q_3, as shown in the example.

EXAMPLE 2.1

Catherine is a junior reporter at a local newspaper. As part of an investigation into consumer affairs she purchases 0.5 kg of lean mince from 12 shops and supermarkets in the town. The resulting data, put into rank order, are as follows:

£1.39 £1.39 £1.46 £1.48 £1.48 £1.50 £1.52 £1.54 £1.60 £1.65 £1.68 £1.72

Find Q_1, Q_2 and Q_3.

Solution

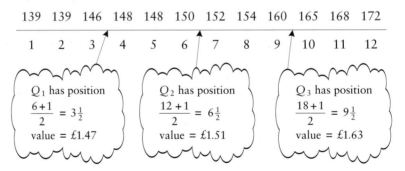

In fact, the upper quartile has a value of £1.625 but this has been rounded up to the nearest penny.

You may encounter different formulae for finding the lower and upper quartiles. The ones given here are relatively easy to calculate and usually lead to values of Q_1 and Q_3 which are close to the true values.

Interquartile range or quartile spread

The difference between the lower and upper quartiles is known as the *interquartile range* or *quartile spread*.

$$\text{Interquartile range } (IQR) = Q_3 - Q_1.$$

In Example 2.1 $IQR + 163 - 147 = 16p$.

The interquartile range covers the middle 50% of the data. It is relatively easy to calculate and is a useful measure of spread as it avoids extreme values. It is said to be resistant to outliers.

Box and whisker plots (boxplots)

The three quartiles and the two extreme values of a data set may be illustrated in a *box and whisker plot*. This is designed to give an easy-to-read representation of the location and spread of a distribution. Figure 2.21 shows a box and whisker plot for the data in Example 2.1.

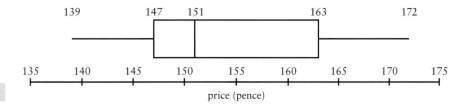

FIGURE 2.21

The box represents the middle 50% of the distribution and the whiskers stretch out to the extreme values.

Figure 2.22 shows a box and whisker plot for the data relating to the ages of the cyclists involved in accidents given on page 7.

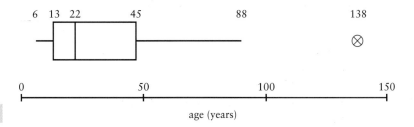

FIGURE 2.22

From the diagram you can see that the distribution has positive or right skewness. The ⊗ indicates an outlier. Outliers are usually labelled as they are often of special interest. The whiskers are drawn to the most extreme data points.

OUTLIERS

Data which are at least $1.5 \times IQR$ beyond the upper or lower quartile may be outliers. Extreme outliers are at least $3 \times IQR$ beyond the upper or lower quartile.

CUMULATIVE FREQUENCY CURVES

When working with large data sets or grouped data, percentiles and quartiles can be found from *cumulative frequency curves* as shown in the next section.

A university student decided to collect data about the cost of books. She had been told that most textbooks cost over £16, and so she took a large sample of 470 textbooks and the results are summarised in the table.

Cost, C (£)	Frequency (no. of books)
$C < 10$	13
$10 \leqslant C < 15$	53
$15 \leqslant C < 20$	97
$20 \leqslant C < 25$	145
$25 \leqslant C < 30$	81
$30 \leqslant C < 35$	40
$35 \leqslant C < 40$	23
$40 \leqslant C < 45$	12
$45 \leqslant C < 50$	6

She decided to estimate the median and the upper and lower quartiles of the costs of the books. (Without the original data you cannot find the actual values so all calculations will be estimates.) The first step is to make a cumulative frequency table, then to plot a cumulative frequency curve.

Cost, C (£)	Frequency	Cost	Cumulative frequency	
$0 \leqslant C < 10$	13	$C < 10$	13	
$10 \leqslant C < 15$	53	$C < 15$	66	See Note 1
$15 \leqslant C < 20$	97	$C < 20$	163	See Note 2
$20 \leqslant C < 25$	145	$C < 25$	308	
$25 \leqslant C < 30$	81	$C < 30$	389	
$30 \leqslant C < 35$	40	$C < 35$	429	
$35 \leqslant C < 40$	23	$C < 40$	452	
$40 \leqslant C < 45$	12	$C < 45$	464	
$45 \leqslant C < 50$	6	$C < 50$	470	

Cumulative frequency curves

Notes

1 Notice that the interval C < 15 means 0 ≤ C < 15 and so includes the 13 books in the interval 0 ≤ C < 10 and the 53 books in the interval 10 ≤ C < 15, giving 66 books in total.
2 Similarly, to find the total for the interval C < 20 you must add the number of books in the interval 15 ≤ C < 20 to your previous total, giving you 66 + 97 = 163.

A cumulative frequency curve is obtained by plotting the *upper boundary* of each class against the cumulative frequency. The points are joined by a smooth curve, as shown in figure 2.23.

In this example the actual values are unknown and the median must therefore be an estimate. It is usual in such cases to find the estimated value of the $\frac{n}{2}$th item. This gives a better estimate of the median than is obtained by using $\frac{n+1}{2}$, which is used for ungrouped data. Similarly, estimates of the lower and upper quartiles are found from the $\frac{n}{4}$th and $\frac{3n}{4}$th items.

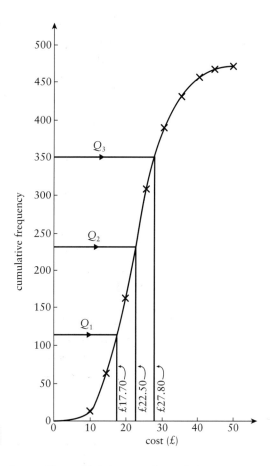

Figure 2.23

The 235th ($\frac{470}{2}$) item of data identifies the median which has a value of about £22.50. The 117.5th ($\frac{470}{4}$) item of data identifies the lower quartile, which has a value of about £17.70 and the 352.5th ($\frac{3}{4} \times 470$) item of data identifies the upper quartile, which has a value of about £27.70.

Notice the distinctive shape of the cumulative frequency curve. It is like a stretched out S-shape leaning forwards.

What about the claim that the majority of textbooks cost more than £16? Q_1 = £17.70. By definition 75% of books are more expensive than this, so this claim seems to be well founded. We need to check exactly how many books are estimated to be more expensive than £16.

From the cumulative frequency curve 85 books cost £16 or less. So 385 books or about 82% are more expensive.

You should be cautious about any conclusions you draw. This example deals with books, many of which have prices like £9.95 or £39.99. In using a cumulative frequency curve you are assuming an even spread of data throughout the intervals and this may not always be the case.

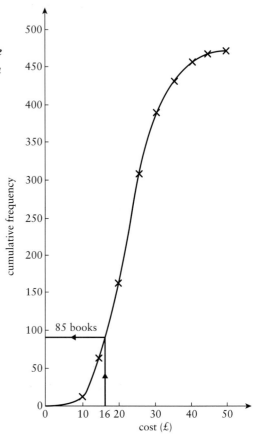

FIGURE 2.24

PERCENTILES

The median is sometimes called the 50th percentile, since 50% of the distribution lies below it.

Similarly

Q_1 = lower quartile = 25th percentile
Q_3 = upper quartile = 75th percentile

The values of percentiles may be read off from cumulative frequency diagrams.

In the example about book prices, suppose we want to know the prices of the most expensive 10% of the books. Then 90% cost less than this figure, £x say, so we find 90% of 470 = 423 books. Then, using the graph, you can read off that the most expensive 10% of the books cost between £34 and £50.

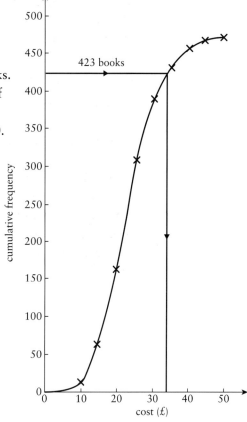

FIGURE 2.25

BOX AND WHISKER PLOTS FOR GROUPED DATA

It is often helpful to draw a box and whisker plot. In cases such as the above when the extreme values are unknown the whiskers are drawn out to the 10th and 90th percentiles. Arrows indicate that the minimum and maximum values are further out.

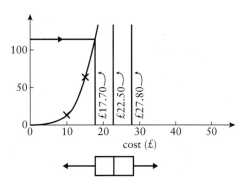

FIGURE 2.26

Exercise 2D

1. For each of the following sets of figures, find
   (i) the median
   (ii) the upper and lower quartiles
   (iii) the interquartile range.

   (a) 6 8 3 2 1 5 4 6 8 5 6 7 8 8 6 6
   (b) 12 5 17 11 4 10 12 19 12 5 9 15 11 16 8
       18 12 8 9 11 12 14 8 14 7
   (c) 25 28 29 30 20 23 23 27 25 28
   (d) 115 123 132 109 127 116 128 132 114 109
       125 134 121 117 118 117 116 123 105 125
   (e) 2 8 4 6 3 5 1 8 2 5 8 0 3 7 8 5
   (f) 12 18 14 16 13 15 11 18 12 15 18 10 13 17 18 15
   (g) 272 278 274 276 273 275 271 278 272 275 278 270 273 277 278 275
   (h) 20 80 40 60 30 50 10 80 20 50 80 0 30 70 80 50

2. Find
   (a) the median
   (b) the upper and lower quartiles
   (c) the interquartile range
   for the scores of golfers in the first round of a competition.

70	I
71	II
72	IIII
73	⊬H III
74	⊬H ⊬H II
75	⊬H II
76	⊬H
77	⊬H I
78	
79	III
80	I
81	
82	I

   (d) Illustrate the data with a box and whisker plot.
   (e) The scores for the second round are illustrated on the box and whisker plot below. Compare the two and say why you think the differences might have arisen.

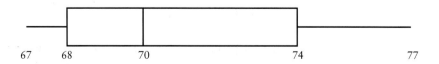

3 The number of goals scored by a hockey team in its matches one season are illustrated on the vertical line chart below.

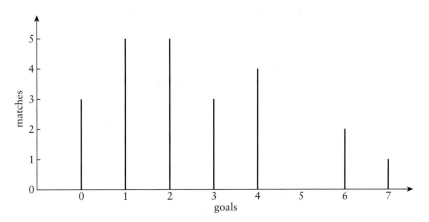

(a) Draw a box and whisker plot to illustrate the same data.
(b) State, with reasons, which you think is the better method of display in this case.

4 One year the yields, y, of a number of walnut trees were recorded to the nearest kilogram as follows:

Yield, y (kg)	Frequency
40–49	1
50–59	5
60–69	7
70–79	4
80–89	2
90–99	1

(a) Construct the cumulative frequency table for these data.
(b) Draw the cumulative frequency graph.
(c) Use your graph to estimate the median and interquartile range of the yields.
(d) Draw a box and whisker plot to illustrate the data.

The piece of paper where the actual figures had been recorded was then found, and these were:

44 59 67 76 52   62 68 78 53 63   69 82 53 65 70
85 93 56 65 74

(e) Use these data to find the median and interquartile range and compare your answers with those you obtained from the grouped data.
(f) What are the advantages and disadvantages of grouping data?

5 The status of the 120 full-time employees of a small factory is linked to their pay.

Status	Pay (£P/week)
Unskilled	$120 \leqslant P < 200$
Skilled	$200 \leqslant P < 280$
Staff	$280 \leqslant P < 400$
Management	$400 \leqslant P < 1000$

The company's personnel department illustrate the numbers in the various groups on this accurately drawn pie chart.

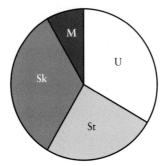

(a) Copy and complete this table, finding the frequencies from the pie chart.

Status	Interval (pay £P/week)	Frequency (employees)	Interval length ($\times £10$)	Frequency density (employees/£10)
Unskilled	$120 \leqslant P < 200$	40	8	5.0
Skilled	$200 \leqslant P < 280$			
Staff	$280 \leqslant P < 400$			
Management	$400 \leqslant P < 1000$			

(b) Hence draw a histogram to illustrate the data.
(c) Is the shape of the histogram about what you would expect?
(d) State, with reasons, which you think is the more appropriate method of display.

6 *UK motorcycle casualties, by age, 1989*

Age (years)	Casualties (%)
16–19	36
20–24	26
25–29	15
30–39	11
40–80	12

(Source: *Social Trends* 21)

(a) Draw a pie chart to illustrate the data, using the five groups given.
(b) Write down the ages of the youngest and oldest person who would come within the interval written as 16–19.
(c) Draw a histogram to represent the data.
(d) Comment on the information in the table, stating with reasons what you consider to be the most and the least dangerous ages for motorcyclists. Do you have enough information to answer this as fully as you would wish?
(e) State, with reasons, which you find the more helpful method of display, the pie chart or the histogram.

Exercise 2E

Examination style questions

1. In a survey of the part-time earnings of some sixth form students the median wage was £24.50. The 75th percentile was £35 and the interquartile range was £23.50.

    Use the quartiles to describe the skewness of the distribution.

    [Edexcel]

2. The following marks were obtained on an A Level mathematics paper by the candidates at one centre.

    ```
    26 54 50  37 54    34 34 66 44 76    45 71   51 75 30
    29 52 43  66 59    22 74 51 49 39    32 37   57 37 18
    54 17 26  40 69    80 90 95 96 95    70 68   97 87 68
    77 76 30 100 98    44 60 46 97 75    52 82   92 51 44
    73 87 49  90 53    45 40 61 66 94    62 39  100 91 66
    35 56 36  74 25    70 69 67 48 65    55 64
    ```

    Draw a stem and leaf diagram to illustrate these marks and comment on their distribution.

3. The ages of a sample of 40 hang-gliders (in years) are given below.

    ```
    28 19 24 20 28    26 22 19 37 40    19 25 65 34 66
    35 69 65 26 17    22 26 45 58 30    31 58 26 29 23
    72 23 21 30 28    65 21 67 23 57
    ```

    (a) Using intervals of ten years, draw a stem and leaf diagram to illustrate these figures.
    (b) Comment on and give a possible explanation for the shape of the distribution.

4 A restaurant owner is concerned about the amount of time customers have to wait before being served. He collects data on the waiting times, to the nearest minute, of 20 customers. These data are listed below.

$$15 \quad 14 \quad 16 \quad 15 \quad 17 \quad 16 \quad 15 \quad 14 \quad 15 \quad 16$$
$$17 \quad 16 \quad 15 \quad 14 \quad 16 \quad 17 \quad 15 \quad 25 \quad 18 \quad 16$$

(a) Find the median and interquartile range of the waiting times.

An outlier is an observation that falls beyond either $1.5 \times$ (interquartile range) above the upper quartile or $1.5 \times$ (interquartile range) below the lower quartile.

(b) Draw a boxplot to represent these data, clearly indicating any outliers.
(c) Find the mean of these data.
(d) Comment on the skewness of the data. Justify your answer.

[Edexcel]

5 The following stem and leaf diagram shows the aptitude scores x obtained by all the applicants for a particular job.

```
Aptitude score              3│1 means 31
3│1 2 9                          (3)
4│2 4 6 8 9                      (5)
5│1 3 3 5 6 7 9                  (7)
6│0 1 3 3 3 5 6 8 8 9            (10)
7│1 2 2 2 4 5 5 5 6 8 8 8 8 9    (14)
8│0 1 2 3 5 8 8 9                (8)
9│0 1 2                          (3)
```

(a) Write down the modal aptitude score.
(b) Find the three quartiles for these data.
(c) On graph paper, draw a boxplot to represent these data.
(d) Explain whether the data are positively skewed or negatively skewed.

[Edexcel, adapted]

EXERCISE 2E

6 At a plant breeding institute, two different strains of a certain species of plant were compared by measuring the length, to the nearest millimetre, of each of 100 leaves of each strain. The results recorded are shown in the table.

Length of leaf (mm)	Frequency Strain A	Frequency Strain B	Length of leaf (mm)	Frequency Strain A	Frequency Strain B
Under 10	3	1	30–34	11	25
10–14	6	4	35–39	6	20
15–19	11	6	40–44	4	11
20–24	22	10	45–49	1	4
25–29	35	16	Over 50	1	3

(a) Draw up a cumulative frequency table for each set of results. On the same axes draw the corresponding cumulative frequency graphs.
(Scales: 2 cm ≡ 10 mm horizontally, 2 cm ≡ 10 leaves vertically.)

(b) Use your graphs to deduce which strain should be developed
  (i) to produce plants with longer leaves
  (ii) to produce plants with uniformity of length of leaf.
Give numerical readings taken from your graphs to support your conclusions.

[MEI, part]

7 *Gross weekly earnings of adults in full-time employment, April 1990 (£)*

	Manual Men	Manual Women	Non-manual Men	Non-manual Women
Upper quartile	280	171	414	264
Median	221	137	312	191
Lower quartile	174	112	231	147

(Source: *New Earnings Survey*, 1990)

(a) Look at these figures and state what you conclude from them.
(b) Using the interquartile range as the measure, compare the spread of the earnings of women with that of men, both manual and non-manual.

8 A teacher recorded the time, to the nearest minute, spent reading during a particular day by each child in a group. The times were summarised in a grouped frequency distribution and represented by a histogram. The first class in the grouped frequency distribution was 10–19 and its associated frequency was eight children. On the histogram the height of the rectangle representing that class was 2.4 cm and the width was 2 cm. The total area under the histogram was 53.4 cm^2.

Find the number of children in the group.

[Edexcel]

9 The following stem and leaf diagram summarises the blood glucose level, in mmol l^{-1}, of a patient, measured daily over a period of time.

```
Blood glucose level    5|0 means 5.0     Totals
5  0 0 1 1 1 2 2 3 3 3 4 4              (12)
5  5 5 6 6 7 8 8 9 9                    ( 9)
6  0 1 1 1 2 3 4 4 4 4                  (10)
6  5 5 6 7 8 9 9                        (  )
7  1 1 2 2 2 3                          (  )
7  5 7 9 9                              (  )
8  1 1 1 2 2 3 3 4                      (  )
8  7 9 9                                ( 3)
9  0 1 1 2                              ( 4)
9  5 7 9                                ( 3)
```

(a) Write down the numbers required to complete the stem and leaf diagram.
(b) Find the median and quartiles of these data.
(c) On graph paper, construct a box plot to represent these data. Show your scale clearly.
(d) Comment on the skewness of the distribution.

[Edexcel]

10 In a certain cross country running competition the times that each of the 136 runners took to complete the course were recorded to the nearest minute. The winner completed the course in 23 minutes and the final runner came in with a time of 78 minutes. The full results are summarised in the table below.

Recorded time	20–29	30–39	40–49	50–59	60–69	70–79
Frequency	7	21	42	37	20	9

(a) Use linear interpolation to estimate the median time.

The upper and lower quartiles of the time taken are 58.1 and 40.9 respectively.

(b) On graph paper, draw a box and whisker plot for the results from this competition. You should mark the end points, the median and the quartiles clearly on your diagram.
(c) Comment on the skewness.

[Edexcel, part]

KEY POINTS

1. *Categorical data* are non-numerical; *discrete data* can be listed; *continuous data* can be measured to any degree of accuracy and it is not possible to list all values.

2. An item of data x may be identified as an *outlier* if x is $1.5 \times IQR$ beyond the upper or lower quartile or if $|x - \bar{x}| > 2 \times$ standard deviation (i.e. if x is more than two standard deviations above or below the sample mean).

3. *Stem and leaf diagrams* (or stemplots) are suitable for discrete or continuous data. All data values are retained as well as indicating properties of the distribution.

4. Bar charts:
   - commonly used to illustrate categorical data
   - vertical axis labelled *frequency*
   - bars usually not touching.

5. Vertical line graphs:
   - commonly used to illustrate discrete data
   - vertical axis labelled *frequency*.

6. Pie charts:
   - total frequency is proportional to area.

7. Histograms:
   - commonly used to illustrate continuous data
   - horizontal axis shows the variable being measured (cm, kg, etc.)
   - vertical axis labelled with the appropriate *frequency density* (per 10 cm, per 100 kg, etc.)
   - no gaps between columns
   - the *frequency density* is *proportional* to the *area* of each column.

8. For a small data set with n items of data,
   - the median, Q_2, is the value of the $\frac{n+1}{2}$th item of data.

   If $\frac{n+1}{2}$ is a whole number, m,
   - the lower quartile, Q_1, is the value of the $\frac{m+1}{2}$th item of data
   - the upper quartile, Q_3, is the value of the $m + \frac{m+1}{2}$th item of data.

   If m is not a whole number ignore the fraction part.

9. Interquartile range $(IQR) = Q_3 - Q_1$.

10. When data are illustrated using a cumulative frequency curve the median, lower and upper quartiles are estimated by identifying the data values with cumulative frequencies $\frac{n}{2}$, $\frac{n}{4}$ and $\frac{3n}{4}$.

11. A box and whisker plot is a useful way of summarising data and showing the median, upper and lower quartiles and any outliers.

Chapter three

SUMMARY OF DATA

A judicious man looks at statistics, not to get knowledge but to save himself from having ignorance foisted on him.

Thomas Carlyle

MEASURES OF CENTRAL TENDENCY

In statistics, it is important for you to be precise about the *average* to which you are referring. Before looking at the different types of average or *measure of central tendency*, you need to be familiar with some notation.

Σ NOTATION AND THE MEAN, $\bar{x}$

A sample of size n taken from a population can be identified as follows:

The first item can be called x_1, the second item x_2 and so on up to x_n.

The sum of these n items of data is given by $x_1 + x_2 + x_3 + \cdots + x_n$.

A shorthand for this is $\sum_{i=1}^{i=n} x_i$ or $\sum_{i=1}^{n} x_i$. This is read as 'the sum of all the terms x_i when i equals 1 to n'.

So $\sum_{i=1}^{n} x_i = x_1 + x_2 + x_3 + \cdots + x_n.$

> $\sum$ is the Greek letter, sigma

If there is no ambiguity about the number of items of data, the subscripts i can be dropped and $\sum_{i=1}^{n} x_i$ becomes $\sum x$.

$\sum x$ is read as 'sigma x' meaning 'the sum of all the x items'.

The *mean* of these n items of data is written as $\bar{x} = \dfrac{x_1 + x_2 + x_3 + \cdots + x_n}{n}$

where $\bar{x}$ is the symbol for the mean, referred to as 'x-bar'.

It is usual to write $\bar{x} = \dfrac{\sum x}{n}$ or $\bar{x} = \dfrac{1}{n}\sum x$.

This is a formal way of writing 'To get the mean you add up all the data values and divide by the total number of data values'.

Often data are presented in a frequency table. The notation for the mean is slightly different in such cases.

Alex is a member of the local bird-watching group. The group are concerned about the effect of pollution and climatic change on the well-being of birds.
One spring Alex surveyed a sample of woodlands and domestic garden sites for Blue Tit nests. Blue Tits usually lay 1–5 eggs. Alex collected data from 50 nests. His data are shown in the following frequency table.

Number of eggs, x	Frequency, f
1	4
2	12
3	9
4	18
5	7
	$\sum f = 50$

This represents 'the sum of the separate frequencies is 50'. That is $4 + 12 + 9 + 18 + 7 = 50$

It would be possible to write out the data set in full as $1, 1, 1, \ldots, 5, 5$ and then calculate the mean as before. However, it would not be sensible and in practice the mean is calculated as follows:

$$\bar{x} = \frac{1 \times 4 + 2 \times 12 + 3 \times 9 + 4 \times 18 + 5 \times 7}{50}$$

$$= \frac{162}{50} = 3.24$$

This represents the sum of each of the x terms multiplied by its frequency

In general, this is written as $\bar{x} = \frac{\sum xf}{n}$

$n = \sum f$

Note

The mean is the most commonly used average in statistics. The mean described here is correctly called the *arithmetic mean*; there are other forms, for example, the geometric mean, harmonic mean and weighted mean, all of which have particular applications.

The mean is used when the total quantity is also of interest. For example, the staff at the water treatment works for a city would be interested in the mean amount of water used per household ($\bar{x}$) but would also need to know the total amount of water used in the city ($\sum x$). The mean can give a misleading result if exceptionally large or exceptionally small values occur in the data set.

There are two other commonly used statistical measures of a typical (or representative) value of a data set. These are the median and the mode.

MEDIAN

The *median* is the value of the middle item when all the data items are ranked in order. If there are n items of data then the median is the value of the $\frac{n+1}{2}$ th item.

If n is odd then there is a middle value and this is the median. In the survey of the cyclists in Chapter 2 we had

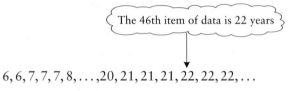

$$6, 6, 7, 7, 7, 8, \ldots, 20, 21, 21, 21, 22, 22, 22, \ldots$$

So for the ages of the 91 cyclists, the median is the age of the $\frac{91+1}{2}$ = 46th person and this is 22 years.

If n is even and the two middle values are a and b then the median is $\frac{a+b}{2}$.
For example, if the reporter had not noticed that 138 was invalid there would have been 92 items of data. Then the median age for the cyclists would be found as follows.

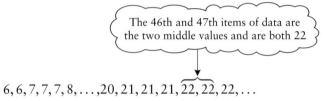

$$6, 6, 7, 7, 7, 8, \ldots, 20, 21, 21, 21, \overbrace{22, 22}, 22, \ldots$$

So the median age for the cyclists is given as the mean of the 46th and 47th items of data. That is, $\frac{22+22}{2}$ = 22.

It is a coincidence that the median turns out to be the same. However, what is important to notice is that an extreme value has little or no effect on the value of the median. The median is said to be resistant to outliers.

The *median* is easy to work out if the data are already ranked, otherwise it can be tedious. However, with the increased availability of computers, it is easier to sort data and so the use of the median is increasing. Illustrating data on a stem and leaf diagram orders the data and makes it easy to identify the median.
The median usually provides a good representative value and, as seen above, it is not affected by extreme values. It is particularly useful if some values are missing; for example, if 50 people took part in a cross country race then the median is halfway between the 25th and 26th values. If some people failed to complete the course the mean would be impossible to calculate, but the median is easy to find.

In finding an *average* salary the median is often a more appropriate measure than the mean since a few people earning very large salaries would have a big effect on the mean but not on the median.

EXERCISE 3A

Mode

The *mode* is the value which occurs most frequently. If two non-adjacent values occur more frequently than the rest, the distribution is said to be *bimodal*, even if the frequencies are not the same for both modes.

Bimodal data usually indicates that the sample has been taken from two populations. For example, a sample of students' heights (male and female) would probably be bimodal, reflecting the different average heights of males and females.

For a small set of discrete data the mode can often be misleading, especially if there are many values the data can take. Several items of data can happen to fall on a particular value. The mode is used when the most probable or most frequently occurring value is of interest. For example, a dress shop manager who is considering stocking a new style would first buy dresses of the new style in the modal size, as she would be most likely to sell those ones.

Which average you use will depend on the particular data you have and on what you are trying to find out.

The measures for the cyclists' ages are summarised below.

Mean	29.9 years
Mode	9 years
Median	22 years

EXERCISE 3A

1 Find the mode, mean and median of these figures:
   (a) 23 46 45 45 29   51 36 41 37 47   45 44 41 31 33
   (b) 110 111 116 119 129   126 132 116 122 130
        116 132 118 122 127   132 126 138 117 111
   (c) 5 7 7 9 1   2 3 5 6 6   8 6 5 7 9   2 2 5 6 6
        6 4 7 7 6   1 3 3 5 7   8 2 8 7 6   5 4 3 6 7

2 For each of these sets of data
   (i) find the mode, mean and median
   (ii) state, with reasons, which you consider to be the most appropriate form of average to describe the distribution.
   (a) The ages of students in a class in years and months:
       14.1 14.11 14.5 14.6 14.0   14.7 14.7 14.9 14.1 14.2
       14.6 14.5 14.8 14.2 14.0   14.9 14.2 14.8 14.11 14.8
       15.0 14.7 14.8 14.9 14.3   14.5 14.4 14.3 14.6 14.1
   (b) The shoe sizes of children in a class:
       3 $2\frac{1}{2}$ 5 4 $3\frac{1}{2}$ 4 4 4 $2\frac{1}{2}$ 6 $4\frac{1}{2}$ 5 $5\frac{1}{2}$ 4 $3\frac{1}{2}$
       $1\frac{1}{2}$ 3 3 4 $2\frac{1}{2}$ $3\frac{1}{2}$ 5 4 3 $4\frac{1}{2}$ $3\frac{1}{2}$ 4 5 $3\frac{1}{2}$ $4\frac{1}{2}$

(c) The numbers of pints of beer drunk by people in *The Crown and Anchor* one Friday evening:

4 0 0 0 $\frac{1}{2}$ 5 3 4 0 0   1$\frac{1}{2}$ 0 4 8 0
4 4 $\frac{1}{2}$ 0 6   3 3 4 5 4 $\frac{1}{2}$ 3 0 4 4

(d) Students' marks on an examination paper:

55 78 45 54 0   62 43 56 71 65   0 67 75 51 100
39 45 66 71 52   71 0 0 59 61   56 59 64 57 63

(e) The scores of a cricketer during a season's matches:

10 23 65 0 1   24 47 2 21 53   5 4 23 169 21
17 34 33 21 0   10 78 1 56 3   2 0 128 12 19

(f) Scores when a die is thrown 40 times:

2 4 5 5 1   3 4 6 2 5   2 4 6 1 2   5 4 4 1 1
3 4 6 5 5   2 3 3 1 6   5 4 2 1 3   3 2 1 6 6

FREQUENCY DISTRIBUTIONS

You will often have to deal with data that are presented in a frequency table. Frequency tables summarise the data and also allow you to get an idea of the shape of the distribution.

EXAMPLE 3.1

Claire runs a fairground stall. She has designed a game where customers pay £1 and are given 10 marbles which they have to try to get into a container 4 metres away. If they get more than 8 in the container they win £5. Before introducing the game to the customers she tries it out on a sample of 50 people. The number of successes scored by each person is noted.

5 7 8 7 5 4 0 9 10 6
4 8 8 9 5 6 3 2 4 4
6 5 5 7 6 7 5 6 9 2
7 7 6 3 5 5 6 9 8 7
5 2 1 6 8 5 4 4 3 3

The data are discrete. They have not been organised in any way, so they are referred to as raw data

Find
(a) the mode
(b) the median
(c) the mean
of the 50 values.
(d) Comment on your results.

Frequency distributions

Solution The *frequency distribution* of these data can be illustrated in a table. The numbers of 0s, 1s, 2s, etc. are counted to give the frequency of each score.

Score	Frequency
0	1
1	1
2	3
3	4
4	6
5	10
6	8
7	7
8	5
9	4
10	1
Total	50

With the data presented in this form it is easier to find or calculate the different averages

(a) The mode score is 5 (frequency 10).

(b) As the number of items of data is even, the distribution has two middle values, the 25th and 26th scores. From the distribution, by adding up the frequencies, it can be seen that the 25th score is 5 and the 26th score is 6. Consequently the median score is $\frac{1}{2}(5 + 6) = 5.5$.

(c) Representing a score by x and its frequency by f, the calculation of the mean is shown in this table.

Score, x	Frequency, f	$x \times f$
0	1	$0 \times 1 = 0$
1	1	$1 \times 1 = 1$
2	3	$2 \times 3 = 6$
3	4	12
4	6	24
5	10	50
6	8	48
7	7	49
8	5	40
9	4	36
10	1	10
	$\sum f = 50$	$\sum xf = 276$

So $\bar{x} = \dfrac{\sum xf}{n}$

$= \dfrac{276}{50} = 5.52$

(d) The values of the mode (5), the median (5.5) and the mean (5.52) are close. This is because the distribution of scores does not have any extreme values and is reasonably symmetrical.

EXERCISE 3B

1 A bag contained six counters numbered 1, 2, 3, 4, 5 and 6. A counter was drawn from the bag, its number was noted and then it was returned to the bag. This was repeated 100 times. The results were recorded in a table giving the frequency distribution shown.

Number, x	Frequency, f
1	15
2	25
3	16
4	20
5	13
6	11

(a) State the mode.
(b) Find the median.
(c) Calculate the mean.

2 A sample of 50 boxes of matches with stated contents 40 matches was taken. The actual number of matches in each box was recorded. The resulting frequency distribution is shown in the table.

Number of matches, x	Frequency, f
37	5
38	5
39	10
40	8
41	7
42	6
43	5
44	4

(a) State the mode.
(b) Find the median.

(c) Calculate the mean.
(d) State, with reasons, which you think is the most appropriate form of average to describe the distribution.

3 A survey of the number of students in 80 classrooms in a school was carried out. The data were recorded in a table as follows.

Number of students, x	Frequency, f
5	1
11	1
15	6
16	9
17	12
18	16
19	18
20	13
21	3
22	1
	$\sum f = 80$

(a) State the mode.
(b) Find the median.
(c) Calculate the mean.
(d) State, with reasons, which you think is the most appropriate form of average to describe the distribution.

4 The tally below gives the scores of the football teams in the matches of the 1982 World Cup finals.

```
 0    JHT JHT JHT JHT JHT JHT I
 1    JHT JHT JHT JHT JHT JHT JHT III
 2    JHT JHT JHT I
 3    JHT III
 4    JHT I
 5    II
 6
 7
 8
 9
10    I
```

(a) Find the mode, mean and median of these data.
(b) State which of these you think is the most representative measure.

5 The vertical line chart below shows the number of times the various members of a school year had to take their driving test before passing it.

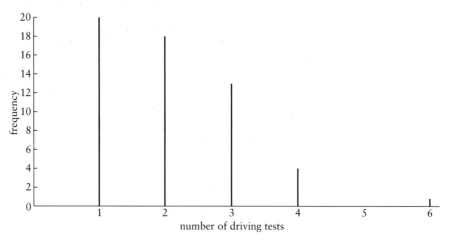

(a) Find the mode, mean and median of these data.

(b) State which of these you think is the most representative measure.

GROUPED DATA

Grouping means putting the data into a number of classes. The number of data items falling into any class is called the *frequency* for that class.

When numerical data are grouped, each item of data falls within a *class interval* lying between *class boundaries*.

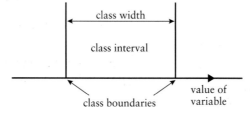

FIGURE 3.1

You must always be careful about the choice of class boundaries because it must be absolutely clear to which class any item belongs. A form with the following wording:

How old are you? Please tick one box.

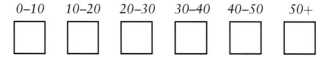

would cause problems. A ten-year-old could tick either of the first two boxes.

A better form of wording would be:

How old are you (in completed years)? Please tick one box.

0–9   10–19   20–29   30–39   40–49   50+
☐   ☐   ☐   ☐   ☐   ☐

Notice that this says 'in completed years'. Otherwise a $9\frac{1}{2}$-year-old might not know which of the first two boxes to tick.

Another way of writing this is:

$$0 \leqslant A < 10 \quad 10 \leqslant A < 20 \quad 20 \leqslant A < 30$$
$$30 \leqslant A < 40 \quad 40 \leqslant A < 50 \quad 50 \leqslant A$$

Even somebody aged 9 years and 364 days would clearly still come in the first group.

WORKING WITH GROUPED DATA

There is often a good reason for grouping raw data.

- There may be a lot of data.
- The data may be spread over a wide range.
- Most of the values collected may be different.

Whatever the reason, grouping data should make it easier to analyse and present a summary of findings, whether in a table or in a diagram.

For some *discrete data* it may not be necessary or desirable to group them. For example, a survey of the number of passengers in cars using a busy road is unlikely to produce many integer values outside the range 0 to 4 (not counting the driver). However, there are cases when grouping the data (or perhaps constructing a stem and leaf diagram) is an advantage.

DISCRETE DATA

At various times during one week the number of cars passing a survey point was noted. Each item of data relates to the number of cars passing during a five-minute period. A hundred such periods were surveyed. The data are summarised in the following frequency table.

Number of cars, x	Frequency, f
0–9	5
10–19	8
20–29	13
30–39	20
40–49	22
50–59	21
60–70	11
	$\sum f = 100$

From the frequency table you can see there is a slight negative (or left) skew.

Estimating the mean

When data are grouped the individual values are lost. This is not often a serious problem; as long as the data are reasonably distributed throughout each interval it is possible to *estimate* statistics such as the mean, knowing that your answers will be reasonably accurate.

To estimate the mean you first assume that all the values in an interval are equally spaced about a mid-point. The mid-points are taken as representative values of the intervals.

The mid-value for the interval 0–9 is $\frac{0+9}{2} = 4.5$.

The mid-value for the interval 10–19 is $\frac{10+19}{2} = 14.5$, and so on.

The $x \times f$ column can now be added to the frequency distribution table and an estimate for the mean found.

Number of cars, x (mid-values)	Frequency, f	$x \times f$
4.5	5	$4.5 \times 5 = 22.5$
14.5	8	$14.5 \times 8 = 116.0$
24.5	13	318.5
34.5	20	690.0
44.5	22	979.0
54.5	21	1144.5
65.0	11	715.0
	$\sum f = 100$	$\sum xf = 3985.5$

The mean is given by

$$\bar{x} = \frac{3985.5}{100}$$

$$= 39.855$$

The original raw data, summarised in the above frequency table, are shown below.

```
10  18  68  67  25  62  49  11  12   8
 9  46  53  57  30  63  34  21  68  31
20  16  29  13  31  56   9  34  45  55
35  40  45  48  54  50  34  32  47  60
70  52  21  25  53  41  29  63  43  50
40  48  45  38  51  25  52  55  47  46
46  50   8  25  56  18  20  36  36   9
38  39  53  45  42  42  61  55  30  38
62  47  58  54  59  25  24  53  42  61
18  30  32  45  49  28  31  27  54  38
```

In this form it is impossible to get an overview of the numbers of cars, nor would listing every possible value in a frequency table (0 to 70) be helpful.

However, grouping the data and estimating the mean was not the only option. Constructing a stem and leaf diagram and using it to find the median would have been another possibility.

The data the reporter used in his article on cycling accidents included the distance from home, in metres, of those involved in cycling accidents. In full these were as follows:

```
3000      75  1200  300    50    10   150  1500   250    25
 200    4500    35    60   120   400  2400   140    45     5
1250    3500    30    75   250  1200   250    50   250   450
  15    4000
```

It is clear that there is considerable spread in the data. They are continuous data and the reporter is aware that the values appear to have been rounded but he does not know to what level of precision. Consequently there is no way of reflecting the level of precision in setting the interval boundaries.

The reporter wants to estimate the mean and decides on the following grouping.

Location relative to home	Distance, d, in metres	Distance mid-value, x	Frequency (number of accidents), f	$x \times f$
Very close	$0 \leqslant d < 100$	50	12	600
Close	$100 \leqslant d < 500$	300	11	3300
Not far	$500 \leqslant d < 1500$	1000	3	3000
Quite far	$1500 \leqslant d < 5000$	3250	6	19 500
			$\sum f = 32$	$\sum xf = 26\,400$

$$\bar{x} = \frac{26\,400}{32} = 825\,\text{m}$$

A summary of the measures of central tendency for the original and grouped accident data is given below.

	Raw data		Grouped data
Mean	$25\,785 \div 32 = 806\,\text{m}$		$825\,\text{m}$
Mode	$250\,\text{m}$	Modal group	$0 \leqslant d < 100\,\text{m}$
Median	$\frac{1}{2}(200 + 250) = 225\,\text{m}$		

CONTINUOUS DATA

For a statistics project Robert, a first year university student, collected the heights of 50 female students.

He constructed a frequency table for his project and included the calculations to find an estimate for the mean of his data.

Height, h	Mid-value, x	Frequency, f	xf
$157 < h \leqslant 159$	158	4	632
$159 < h \leqslant 161$	160	11	1760
$161 < h \leqslant 163$	162	19	3078
$163 < h \leqslant 165$	164	8	1312
$165 < h \leqslant 167$	166	5	830
$167 < h \leqslant 169$	168	3	504
		$\sum f = 50$	$\sum xf = 8116$

$$\bar{x} = \frac{8116}{50} = 162.32$$

Working with grouped data

Note

Class boundaries

His tutor was concerned about the class boundaries and asked Robert 'To what degree of accuracy have you recorded your data?' Robert told him 'I rounded all my data to the nearest centimetre'. Robert showed his tutor his raw data.

163	160	167	168	166	164	166	162	163	163
165	163	163	159	159	158	162	163	163	166
164	162	164	160	161	162	162	160	169	162
163	160	167	162	158	161	162	163	165	165
163	163	168	165	165	161	160	161	161	161

Robert's tutor said that the class boundaries should have been

$157.5 \leq h < 159.5$

$159.5 \leq h < 161.5$, and so on.

He explained that a height recorded to the nearest centimetre as 158 cm has a value in the interval 158 ± 0.5 cm (this can be written as $157.5 \leq x < 158.5$). Similarly the actual values of those recorded as 159 cm lie in the interval $158.5 \leq x < 159.5$. So, the interval $157.5 \leq h < 159.5$ covers the actual values of the data items 158 and 159. The interval $159.5 \leq h < 161.5$ covers the actual values of 160 and 161 and so on.

What adjustment does Robert need to make to his estimated mean in the light of his tutor's comments?

You are not always told the level of precision of summarised data and the class widths are not always equal. Also, there are different ways of representing class boundaries, as the following example illustrates.

EXAMPLE 3.2

The frequency distribution shows the lengths of telephone calls made by Emily during August. Choose suitable mid-class values and estimate Emily's mean call time for August.

Solution

Time (seconds)	Frequency, f	Mid-value, x	xf
0–	39	30	1170
60–	15	90	1350
120–	12	150	1800
180–	8	240	1920
300–	4	400	1600
500–1000	1	750	750
	$\sum f = 79$		$\sum xf = 8590$

$$\bar{x} = \frac{8590}{79} = 108.7 \text{ seconds (3 sf)}$$

Notes

1 The interval '0–' can be written as $0 \leqslant x < 60$, the interval '60–' can be written as $60 \leqslant x < 120$, and so on, up to '500–1000', which can be written as $500 \leqslant x \leqslant 1000$.
2 There is no indication of the level of precision of the recorded data. They may have been recorded to the nearest second.
3 The class widths vary.

Exercise 3C

1 A college nurse keeps a record of the heights, measured to the nearest centimetre, of a group of students she treats.

Her data are summarised in the following grouped frequency table.

Height (cm)	110–119	120–129	130–139	140–149	150–159	160–169	170–179	180–189
No. of students	1	3	10	28	65	98	55	15

Choose suitable mid-class values and calculate an estimate for the mean height.

2 A junior school teacher noted the time to the nearest minute a group of children spent reading during a particular day.

The data are summarised as follows:

Time (nearest minute)	Number of children
20–29	12
30–39	21
40–49	36
50–59	24
60–69	12
70–89	9
90–119	2

(a) Choose suitable mid-class values and calculate an estimate for the mean time spent reading by the pupils.
(b) Some time later, the teacher collected similar data from a group of 25 children from a neighbouring school. She calculated the mean to be 75.5 minutes. Compare the estimate you obtained in part (a) with this value.

What assumptions must you make for the comparison to be meaningful?

3 The stated age of the 91 cyclists considered in Chapter 2 is summarised by the following grouped frequency distribution.

Stated age (years)	Frequency
0–9	13
10–19	26
20–29	16
30–39	10
40–49	6
50–59	5
60–69	14
70–79	0
80–89	1
	$\sum f = 91$

(a) Choose suitable mid-interval values and calculate an estimate of the mean stated age.

(b) Make a suitable error adjustment to your answer to part (a) to give an estimate of the mean age of the cyclists.

4 In an agricultural experiment, 320 plants were grown on a plot. The lengths of the stems were measured, to the nearest centimetre, 10 weeks after planting. The lengths were found to be distributed as in the following table:

Length, x (cm)	Frequency (number of plants)
$20.5 \leq x < 32.5$	30
$32.5 \leq x < 38.5$	80
$38.5 \leq x < 44.5$	90
$44.5 \leq x < 50.5$	60
$50.5 \leq x < 68.5$	60

Calculate an estimate of the mean of stem lengths from this experiment.

5 The newspaper reporter considered choosing different classes for the data dealing with the cyclists who were involved in accidents.

He summarised the distances from home of the 91 cyclists as follows:

Distance, d (metres)	Frequency
$0 \leqslant d < 50$	7
$50 \leqslant d < 100$	5
$100 \leqslant d < 150$	2
$150 \leqslant d < 200$	1
$200 \leqslant d < 300$	5
$300 \leqslant d < 500$	3
$500 \leqslant d < 1000$	0
$1000 \leqslant d < 5000$	9
	$\sum f = 32$

(a) Choose suitable class mid-values and estimate the mean.

(b) The mean of the raw data is 806 m and his previous grouping gave an estimate for the mean of 825 m. Compare your answer to this value and comment.

6 A case containing 270 oranges was opened and each orange was weighed. The masses, given to the nearest gram, were grouped and the resulting distribution is as follows:

Mass, x (grams)	Frequency (number of oranges)
60–99	20
100–119	60
120–139	80
140–159	50
160–220	60

(a) State the class boundaries for the interval 60–99.

(b) Calculate an estimate for the mean mass of the oranges from the crate.

MEASURES OF SPREAD

In the last section you saw how an estimate for the mean can be found from grouped data. The mean is just one example of a *typical value* of a data set. You also saw how the mode and the median can be found from small data sets. Chapter 2 considered the use of the median as a *typical value* when dealing with grouped data and also the *interquartile range* as a *measure of spread*. In this chapter we will consider the range, the variance and the standard deviation as measures of spread.

RANGE

The simplest measure of spread is the *range*. This is just the difference between the largest value in the data set (the upper extreme) and the smallest value (the lower extreme).

> *Range = largest − smallest*

The figures below are the prices, in pence, of a 100 g jar of *Nesko* coffee in ten different shops:

    161   161   163   163   167   168   170   172   172   172

The range for this data is

    Range = 172 − 161 = 11p.

EXAMPLE 3.3

Ruth is investigating the amount of money students earn from part-time work on one particular weekend. She collects and orders data from two classes and this is shown below.

Class 1									
10	10	10	10	10	10	12	15	15	15
16	16	16	16	18	18	20	25	38	90

Class 2									
10	10	10	10	10	10	12	12	12	12
15	15	15	15	16	17	18	19	20	20
25	35	35							

She calculates the mean amount earned for each class. Her results are

    Class 1: $\bar{x}_1$ = £19.50
    Class 2: $\bar{x}_2$ = £16.22

She concludes that the students in Class 1 each earn about £3 more, on average, than do the students in Class 2.

Her teacher suggests she look at the spread of the data. What further information does this reveal?

Solution Ruth calculates the range for each class: Range (Class 1) = £80
Range (Class 2) = £25

She concludes that the part-time earnings in Class 1 are much more spread out.

However, when Ruth looks again at the raw data she notices that one student in Class 1 earned £90, considerably more than anybody else. If that item of data is ignored then the spread of data for the two classes is similar.

One of the problems with the range is that it is prone to the effect of extreme values.

The range does not use all of the available information; only the extreme values are used. In quality control this can be an advantage as it is very sensitive to something going wrong on a production line. Also the range is easy to calculate. However, usually we want a measure of spread that uses all the available data and that relates to a central value.

VARIANCE AND STANDARD DEVIATION

A common measure of spread is the standard deviation. Although fiddly to calculate it does play a central role in advanced statistics. The standard deviation does take all the data points into account, unlike the range or the interquartile range.

To find the standard deviation it is helpful to calculate the variance first. There are two ways of doing this:

Method 1
- Find the mean, $\bar{x}$
- Compute the squared deviations, $(x - \bar{x})^2$
- Sum up them up, $\sum (x - \bar{x})^2$
- Find their mean, $\dfrac{\sum (x - \bar{x})^2}{n}$ = variance

- Then standard deviation = $\sqrt{\text{variance}}$

so $$\text{standard deviation} = \sqrt{\dfrac{\sum (x - \bar{x})^2}{n}}$$

So for the data set 6, 7, 10, 17
- Mean $\bar{x} = \dfrac{6 + 7 + 10 + 17}{4} = \dfrac{40}{4} = 10$
- Squared deviations $(6 - 10)^2 = 16$, $(7 - 10)^2 = 9$, $(10 - 10)^2 = 0$, $(17 - 10)^2 = 49$
- Their mean is $\dfrac{16 + 9 + 0 + 49}{4} = \dfrac{74}{4} = 18.5$
- Standard deviation = $\sqrt{18.5} = 4.3$

Method 2
The formula
$$\frac{\sum(x-\bar{x})^2}{n}$$
may be rearranged to give the alternative result
$$\frac{\sum x^2}{n} - \bar{x}^2.$$
This is remembered in words as 'the mean of the squares minus the square of the mean'. It can make the calculation of standard deviation considerably simpler, especially when $\bar{x}$ is not a whole number.

- Find the mean, $\bar{x}$
- Find the squares of the values, x^2
- Sum them up, $\sum x^2$
- Compute $\dfrac{\sum x^2}{n} - \bar{x}^2$ = variance
- Then standard deviation = $\sqrt{\text{variance}}$

So for the data set 6, 7, 10, 17

- Mean $\bar{x} = \dfrac{6+7+10+17}{4} = \dfrac{40}{4} = 10$
- Squared values are 36, 49, 100, 289
- $\sum x^2 = 36 + 49 + 100 + 289 = 474$
- Variance = $\dfrac{474}{4} - 10^2 = 118.5 - 100 = 18.5$
- Standard deviation = $\sqrt{18.5} = 4.3$

In practice you will make extensive use of your calculator's statistical functions to find the mean and standard deviation of sets of data.

Care should be taken as the notations S, s, sd, σ, and $\hat{\sigma}$ are used differently by different calculator manufacturers, authors and users.

In more advanced work, the *sample variance* is often used as an estimate for the *population variance*. When this is the case a slightly different formula is used for the sample variance:
$$\frac{1}{n-1}\sum f_i(x_i - \bar{x})^2.$$

Throughout this book we use sd^2 to mean the variance of the sample considered, where
$$sd^2 = \frac{1}{n}\sum x^2 f - \bar{x}^2.$$

The following examples involve finding or using the sample variance.

EXAMPLE 3.4

The following information relates to a sample of size 60.
$\sum x^2 = 18\,000$, $\sum x = 960$. Find the mean and the standard deviation.

Solution

$$\bar{x} = \frac{\sum x}{n} = \frac{960}{60} = 16$$

$$\text{variance} = \frac{1}{n}\sum x^2 - \bar{x}^2 = \frac{18\,000}{60} - 16^2 = 44$$

standard deviation = $\sqrt{44}$ = 6.63 (3 sf).

EXAMPLE 3.5

The following information relates to a sample of size 60.
$\sum (x - \bar{x})^2 = 2000$, $\sum x = 960$. Find the mean and the standard deviation.

Solution

$$\bar{x} = \frac{960}{60} = 16$$

$$\text{variance} = \frac{1}{n}\sum (x - \bar{x})^2 = \frac{2000}{60} = 33.3\ldots$$

standard deviation = $\sqrt{33.3\ldots}$ = 5.77 (3 sf).

EXAMPLE 3.6

As part of her job as quality controller, Stella collected data relating to the life expectancy of a sample of 60 light bulbs produced by her company. The mean life was 650 hours and the standard deviation was 8 hours. A second sample of 80 bulbs was taken by Sol and resulted in a mean life of 660 hours and standard deviation of 7 hours.

Find the overall mean and standard deviation.

Solution Overall mean:

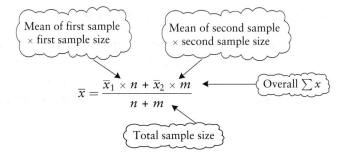

$$\bar{x} = \frac{650 \times 60 + 660 \times 80}{60 + 80}$$

$$= \frac{91\,800}{140} = 655.71\ldots$$

$$= 656 \text{ hours (3 sf)}.$$

For Stella's sample the variance is 8^2. Therefore $8^2 = \dfrac{\sum x_1^2}{60} - 650^2$.

For Sol's sample the variance is 7^2. Therefore $7^2 = \dfrac{\sum x_2^2}{80} - 660^2$.

From the above Stella found that $\sum x_1^2 = (8^2 + 650^2) \times 60 = 25\,353\,840$ and $\sum x_2^2 = 34\,851\,920$.

The overall variance is — Overall $\sum x^2$

Do not round off numbers until you have completed all calculations

$$\frac{25\,353\,840 + 34\,851\,920}{140} - 655.71\ldots^2$$

The total number of light bulbs is 140

$$= 430\,041.14\ldots - 429\,961.22\ldots$$
$$= 79.91\ldots$$

The overall standard deviation is $\sqrt{79.91\ldots} = 8.94$ hours (3 sf).

THE STANDARD DEVIATION AND OUTLIERS

Data sets may contain extreme values and when this occurs you are faced with the problem of how to deal with them.

Many data sets are samples drawn from parent populations which are normally distributed. In these cases approximately:

- 68% of the values lie within 1 standard deviation of the mean
- 95% lie within 2 standard deviations of the mean
- 99.75% lie within 3 standard deviations of the mean.

You will learn more about the normal distribution in Chapter 7. If a particular value is *more than two standard deviations from the mean* it should be investigated as possibly not belonging to the data set. If it is as much as three standard deviations or more from the mean then the case for investigating it is even stronger.

The 2-standard-deviation test should not be seen as a way of defining outliers. It is only a way of identifying those values which might need investigating.

In an A level German class the examination marks at the end of the year are:

35 52 55 61 96 63 50 58 58 49 61

The value 96 was thought to be significantly greater than the other values. The mean and standard deviation of the data are $\bar{x} = 58$ and $sd = 14.2$. The value 96 is more than two standard deviations above the mean:

FIGURE 3.2

When investigated further it turned out that the mark of 96 was achieved by a German boy whose family had moved to Britain and who had taken A level German because he wanted to study German at university. It might be appropriate to omit this value from the data set.

The times taken, in minutes, for some train journeys between Hereford and Shrewsbury were recorded as shown.

56 61 57 55 58 57 5 60 61 59

It is unnecessary here to calculate the mean and standard deviation. The value 5 minutes is obviously a mistake and should be omitted unless it is possible to correct it by referring to the original source of data.

EXERCISE 3D

1 (a) Find the mean of the following data.

0 0 0 1 1 1 1 2 2 2 2 2 3 3 3 3 4 4 4 4 5 5

(b) Find the standard deviation using both forms of the formula.

2 Find the mean and standard deviation of the following data.

x	3	4	5	6	7	8	9
f	2	5	8	14	9	4	3

3 Steve Race and Roy Bull are football players. In the 30 games played so far this season their scoring record is as follows.

Goals scored	0	1	2	3	4
Frequency (Steve)	12	8	8	1	1
(Roy)	4	21	5	0	0

(a) Find the mean and the standard deviation of the number of goals each player scored.

(b) Comment on the players' goal scoring records.

EXERCISE 3D

4 For a set of 20 items of data $\sum x = 22$ and $\sum x^2 = 55$. Find the mean and the standard deviation of the data.

5 For a data set of 50 items of data $\sum (x - \bar{x})^2 f = 8$ and $\sum xf = 20$. Find the mean and the standard deviation of the data.

6 Two thermostats were used under identical conditions. The water temperatures, in °C, are given below:

Thermostat A: 24 25 27 23 26
Thermostat B: 26 26 23 22 28

(a) Calculate the mean and standard deviation for each set of water temperatures.
(b) Which is the better thermostat? Give a reason.

A second sample of data was collected using thermostat A.

25 24 24 25 26 25 24 24

(c) Find the overall mean and the overall standard deviation for the two sets of data for thermostat A.

7 Mrs Davies has a choice of routes to work. She timed her journey along each route on several occasions and the times in minutes are given below.

Town route: 15 16 20 28 21
Country route: 19 21 20 22 18

(a) Calculate the mean and standard deviation of each set of journey times.
(b) Which route would you recommend? Give a reason.

8 In a certain district, the mean annual rainfall is 80 cm, with standard deviation 4 cm.
(a) One year it was 90 cm. Was this an exceptional year?
(b) The next year had a total of 78 cm. Was that exceptional?

Jake, a local amateur meteorologist, kept a record of the weekly rainfall in his garden. His first data set, comprising 20 weeks of figures, resulted in a mean weekly rainfall of 1.5 cm. The standard deviation was 0.1 cm. His second set of data, over 32 weeks, resulted in a mean of 1.7 cm and a standard deviation of 0.09 cm.
(c) Calculate the overall mean and the overall standard deviation for the whole year.
(d) Estimate the annual rainfall in Jake's garden.

9 A farmer expects to harvest a crop of 3.8 tonnes, on average, from each hectare of his land, with standard deviation 0.2 tonnes.

One year there was much more rain than usual and he harvested 4.1 tonnes per hectare.
  (a) Was this exceptional?
  (b) Do you think the crop was affected by the unusual weather or was the higher yield part of the variability which always occurs?

10 The table shows the number of club ranking points scored by a bridge player in a series of 30 games.

$$6\ 6\ 3\ 8\ 7\ 9\ 8\ 6\ 5\ 0\ 8\ 7\ 6\ 9\ 8$$
$$6\ 7\ 2\ 8\ 7\ 6\ 6\ 8\ 9\ 7\ 4\ 8\ 7\ 5\ 9$$

Use your calculator to find the mean and standard deviation of these scores.

In a further 20 games, data on the player's scores were summarised by $\sum x = 143$, $\sum x^2 = 1071$. Find the mean and standard deviation of the scores in all 50 games.

[MEI]

11 As part of a biology experiment Andrew caught and weighed 120 minnows. He used his calculator to find the mean and standard deviation of their weights:

Mean 26.231 g
Standard deviation 4.023 g

  (a) Find the total weight, $\sum x$, of Andrew's 120 minnows.

  (b) Use the formula

  $$\text{standard deviation} = \sqrt{\frac{\sum x^2}{n} - \bar{x}^2}$$

  to find $\sum x^2$ for Andrew's minnows.

Another member of the class, Sharon, did the same experiment with minnows caught from a different stream. Her results are summarised by:

$$n = 80$$
$$\bar{x} = 25.214$$
$$\text{standard deviation} = 3.841$$

Their teacher says they should combine their results into a single set but they have both thrown away their measurements.
  (c) Find n, $\sum x$ and $\sum x^2$ for the combined data set.
  (d) Find the mean and standard deviation for the combined data set.

LINEAR CODING

Consider the following set of grouped data:

Number, x	Frequency, f
3510	6
3512	4
3514	3
3516	1
3518	2
3520	4

Calculating the mean is quite straightforward, but will involve very large and awkward numbers. A useful shortcut would be to subtract 3510 from all the numbers, calculate the mean of the simpler data, then restore the true value by adding 3510 back on again.

This is how the calculations are done.

Number, x	Number − 3510, y	Frequency, f	$y \times f$
3510	0	6	$0 \times 6 = 0$
3512	2	4	$2 \times 4 = 8$
3514	4	3	$4 \times 3 = 12$
3516	6	1	$6 \times 1 = 6$
3518	8	2	$8 \times 2 = 16$
3520	10	4	$10 \times 4 = 40$
		$\sum f = 20$	$\sum yf = 82$

Average (mean) = $\frac{82}{20}$ = 4.1

3510 is now added back:

3510 + 4.1 = 3514.1

The method above is called *linear coding* and is used for two reasons:

(a) to simplify messy arithmetic
(b) to convert between different units.

Consider the following data on the heights of female students.

Height, h (cm) mid-points	Frequency, f
158.5	4
160.5	11
162.5	19
164.5	8
166.5	5
168.5	3
	$\sum f = 50$

The arithmetic involved in calculating the mean and the standard deviation can be simplified considerably as follows:

The h values are replaced by x values, which are found by

(a) subtracting 158.5 from the h values,

then further simplifying the resulting values, 0, 2, 4, 6, 8 and 10, by

(b) dividing by 2, giving 0, 1, 2, 3, 4 and 5.

Height, h (cm) mid-points	x	Frequency, f	xf	x^2f
158.5	$158.5 - 158.5 = 0 \div 2 = 0$	4	0	0
160.5	$160.5 - 158.5 = 2 \div 2 = 1$	11	11	11
162.5	$162.5 - 158.5 = 4 \div 2 = 2$	19	38	76
164.5	$164.5 - 158.5 = 6 \div 2 = 3$	8	24	72
166.5		4	5	20	80
168.5		5	3	15	75
		$\sum f = 50$	$\sum xf = 108$	$\sum x^2 f = 314$

$$\bar{x} = \frac{108}{50} = 2.16$$

$$(sd_x)^2 = \frac{314}{50} - 2.16^2 = 1.6144$$

$$sd_x = 1.27 \text{ cm}$$

In this example the data has been *coded* as $x = \dfrac{h - 158.5}{2}$. From this, $h = 2x + 158.5$.

$\therefore \quad \bar{h} = 2\bar{x} + 158.5$ and $sd_h = 2sd_x$.

EXERCISE 3E

We can now find the mean and standard deviation of the original data.

$$\bar{h} = 2 \times 2.16 + 158.5 = 162.8 \text{ cm}$$
$$sd_h = 2 \times 1.27 = 2.54 \text{ cm}$$

The following example illustrates how linear coding can be used to convert between units.

EXAMPLE 3.7

For a period of ten days during August the mean temperature in Gresham, Oregon, was 80° Fahrenheit. The standard deviation during that period (in degrees Fahrenheit) was 0.7 °F. Find the mean temperature and the standard deviation in degrees Celsius.

Solution The conversion formula is $c = \dfrac{5(f - 32)}{9}$

$\left(\text{which can be written as } c = \dfrac{5f}{9} - \dfrac{160}{9}\right).$

So
$$\bar{c} = \dfrac{5 \times 80}{9} - \dfrac{160}{9} = \dfrac{240}{9}$$
$$= 26\tfrac{2}{3} \, °C \quad \text{or} \quad 26.7 \, °C$$

Subtracting 32 does not affect the spread so the standard deviation is

$$sd_c = \tfrac{5}{9} \times sd_f = \dfrac{5 \times 0.7}{9} = 0.4 \, °C$$

The coded values are easy to calculate, even without a calculator!

> In general, if the coded value, x, is given by a linear equation (or code) of the form $x = \dfrac{y - a}{b}$, then the original value, y, can be found using the equation $y = a + bx$. And, if $\bar{x}$ and sd_x are the mean and standard deviation of the coded x values, then $\bar{y}$ and sd_y, the mean and the standard deviation of the original data, the y values, can be found using $\bar{y} = a + b\bar{x}$ and $sd_y = bsd_x$.

EXERCISE 3E

1 Calculate the mean and standard deviation of the following masses, measured to the nearest gram, using a suitable system of coding.

Mass (g)	241–244	245–248	249–252	253–256	257–260	261–264
Frequency	4	7	14	15	7	3

2 A production line produces steel bolts which have a nominal length of 95 mm. A sample of 50 bolts is taken and measured to the nearest 0.1 mm. Their deviations from 95 mm are recorded in tenths of a millimetre and summarised as $\sum x = -85$, $\sum x^2 = 734$. (For example, a bolt of length 94.2 mm would be recorded as -8.)
   (a) Find the mean and standard deviation of the x values.
   (b) Find the mean and standard deviation of the lengths of the bolts in millimetres.
   (c) One of the figures recorded is -18. Suggest why this can be regarded as an outlier.
   (d) The figure of -18 is thought to be a mistake in the recording. Calculate the new mean and standard deviation of the lengths *in millimetres*, with the -18 value removed.

3 A survey of households in a particular area reveals that the mean and standard deviation of the number of units of electricity used in a quarter are 853 kWh and 279 kWh respectively. The cost per kWh is 7.2p and the standing charge is £12.56, excluding VAT.
   (a) Find the mean and standard deviation of the cost of the electricity used, excluding VAT.
   (b) Find the mean and standard deviation of the cost of the electricity used, including VAT (at 17.5%).

4 A system is used in a college of further education to predict a student's A level grade in a particular subject using their GCSE results. The GCSE score is g and the A level score is a and for Maths in 1999 the equation of the line of best fit relating them was $a = 2.6g - 9.42$.

This year there are 66 second-year students and their GCSE scores are summarised as $\sum g = 408.6$, $\sum g^2 = 2545.06$.
   (a) Find the mean and standard deviation of the GCSE scores.
   (b) Find the mean and standard deviation of the predicted A level scores using the 1999 line of best fit.

5 (a) Find the mode, mean and median of:

   2 8 6 5 4   5 6 3 6 4   9 1 5 6 5

   Hence write down, without further working, the mode, mean and median of:

   (b) 20 80 60 50 40   50 60 30 60 40   90 10 50 60 50
   (c) 12 18 16 15 14   15 16 13 16 14   19 11 15 16 15
   (d) 4 16 12 10 8   10 12 6 12 8   18 2 10 12 10

EXERCISE 3F

6 A manufacturer produces electrical cable which is sold on reels. The reels are supposed to hold 100 metres of cable. In the quality control department the length of cable on randomly chosen reels is measured. These measurements are recorded as deviations, in centimetres, from 100 m. (So, for example, a length of 99.84 m is recorded as −16.)

For a sample of 20 reels the recorded values, x, are summarised by

$$\sum x = -86 \qquad \sum x^2 = 4281$$

(a) Calculate the mean and standard deviation of the values of x.
(b) Hence find the mean and standard deviation, in metres, of the lengths of cable on the 20 reels.
(c) Later it is noticed that one of the values of x is −47, and it is thought that so large a value is likely to be an error. Give a reason to support this view.
(d) Find the new mean and standard deviation of the values of x when the value −47 is discarded.

[MEI]

EXERCISE 3F **Examination style questions**

1 A recent survey of fee-paying schools found that one city had 11 such schools with the fees per pupil (correct to the nearest £100) being as follows.

£3800   £11 100   £3500   £3700   £3800   £3100
£2800   £3500   £3700   £3900   £3500

(a) Find the mean, median and mode of the data. Comment briefly on any substantial differences between these three measures.
(b) Find the standard deviation of the data.
(c) Explain whether the fee of £11 100 should be regarded as an outlier.

In fact the school charging £11 100 is a small school for musically talented pupils.
(d) Explain why the mean as calculated in part (a) will **not** be the mean fee paid per pupil at the 11 schools.

[MEI]

2 On her summer holiday, Felicity recorded the temperatures at noon each day for use in a statistics project. The values recorded, f degrees Fahrenheit, were as follows, correct to the nearest degree.

47   59   68   62   49   67   66   73   70   68   74   84   80   72

(a) Represent Felicity's data on a stem and leaf diagram. (An unsorted diagram is sufficient.) Comment on the shape of the distribution.
(b) Use your calculator to find the mean and standard deviation of Felicity's data.
(c) The formula for converting temperatures from f degrees Fahrenheit to c degrees Celsius is $c = \frac{5}{9}(f - 32)$. *Use this formula* to estimate the mean and standard deviation of the temperatures in degrees Celsius.

[MEI]

3 In tenpin bowling the player attempts to knock down all ten skittles with one ball. If all ten are knocked down the player's turn ends without a second ball being bowled but if any skittles are left standing the player attempts to knock them down with a second ball. After the second ball the player's turn ends even if some skittles remain standing.

A novice player bowls a total of 34 balls. The numbers of skittles knocked down per ball are as follows.

Number of skittles	0	1	2	3	4	5	6	7	8	9	10
Frequency	6	3	1	7	8	3	2	3	0	1	0

(a) Use your calculator to determine the mean and standard deviation of the numbers of skittles knocked down per ball.

The diagrams below show, for two players A and B, the numbers of skittles knocked down per ball. (No scale is given on the frequency axis, but the data are for many hundreds of balls bowled.)

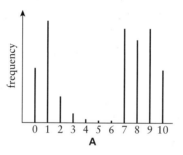
A

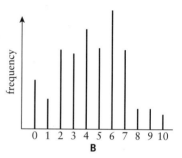
B

(b) Without performing any calculations, **estimate** the mean number of skittles per ball knocked down by each of players A and B.

(c) Without performing any calculations, compare the standard deviations for the two sets of data, explaining your reasoning.

(d) One of the players is a considerably better bowler than the other. State which is which, giving your reasons carefully.

[MEI]

4 In the GCSE year of Old Hall High School there are 207 pupils. The school offers 16 subjects at GCSE. The numbers of pupils taking the various subjects are as follows.

Subject	A	B	C	D	E	F	G	H	I	J	K	L	M	N	O	P
Number	207	207	207	167	165	150	148	121	112	87	45	44	31	25	18	6

(a) What do the data indicate about subjects A, B and C?
(b) Find the mean, median and mode for the number of pupils per subject. Comment briefly on these as 'measures of central tendency'.

EXERCISE 3F

(c) Find the standard deviation of the number of pupils per subject.
(d) Determine the mean number of GCSEs taken per pupil.
(e) Given that there are 79 different GCSE classes (with each pupil attending one class for each subject studied) determine the mean size of a GCSE class.

[MEI]

5 A golf tournament is taking place. For each round, the players' scores are recorded relative to a fixed score of 72. (For example, a true score of 69 would be recorded as −3.)

The recorded scores, x, for the ten players to complete the first round were:

4   −3   −7   6   2   0   0   3   5   2

(a) Calculate the mean and standard deviation of the values of x.
(b) Deduce the mean and standard deviation of the true scores.

In the second round of the tournament, the recorded scores, x, for the same ten golfers produced a mean of −0.3 and standard deviation 2.9.

(c) Comment on how the performance of the golfers has changed from the first to the second round.
(d) Calculate the mean and standard deviation of the twenty true scores for the two rounds.

[MEI]

6 The hourly wages, £x, of the 15 workers in a small factory are as follows:

£6.60   £3.40   £6.45   £5.20   £3.60   £7.25   £9.60   £3.75
£4.20   £8.75   £5.75   £4.50   £3.95   £4.75   £12.25

(a) Illustrate the data in a stem and leaf diagram, using pounds for the stem and pence for the leaves. Clearly indicate the median wage. State the range.
(b) Given that $\sum x = 90.00$ and $\sum x^2 = 631.25$, calculate the mean and standard deviation of hourly wages of the workers.

After delicate wage negotiations, the workers are offered a choice of one of the following pay rises:

(A) an increase of 30p per hour;   (B) a 5% rise in hourly rates.

(c) Use your answers in part **(b)** to deduce the mean and standard deviation of the hourly wages of the 15 workers under both schemes.
(d) Explain why the management would not mind which scheme was implemented, but the workers might.

[MEI]

7 A frequency diagram for a set of data is shown below. No scale is given on the frequency axis, but summary statistics are given for the distribution:

$$\sum f = 50, \quad \sum fx = 100, \quad \sum fx^2 = 344.$$

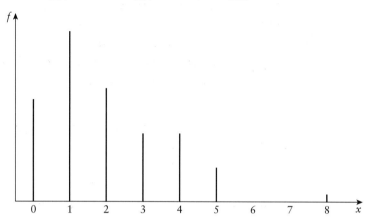

(a) State the mode of the data.
(b) Identify two features of the distribution.
(c) Calculate the mean and standard deviation of the data and explain why the value 8, which occurs just once, may be regarded as an outlier.
(d) Explain how you would treat the outlier if the diagram represents
　(i) the difference of the scores obtained when throwing a pair of ordinary dice
　(ii) the number of children per household in a neighbourhood survey.
(e) Calculate new values for the mean and standard deviation if the single outlier is removed.

[MEI]

8 A farmer gathers apples from his orchard. The apples are graded according to their diameter measured in millimetres.

The diagram shows the distribution of diameters of a sample of the apples. The scale on the vertical axis represents the frequency density (that is, frequency per mm).

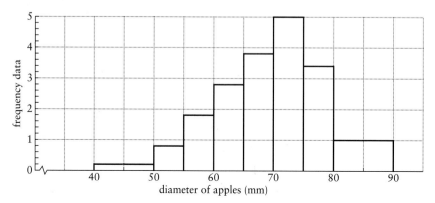

EXERCISE 3F

(a) How many apples are there in the modal class?
(b) Show that the sample contains 100 apples.
(c) Calculate an estimate of the mean diameter of apples in the sample. Explain why your answer is only an estimate.
(d) Find an estimate of the median diameter.
(e) Describe the shape of the distribution.
(f) The standard deviation is approximately 9 mm. State, with reasons, the number of apples you regard as outliers.

[MEI]

9 The table gives the ages, in completed years, of the population in a particular region of the United Kingdom.

Age	0–4	5–15	16–44	45–64	65–79	80 and over
Number (in thousands)	260	543	1727	756	577	135

A histogram of this data was drawn with age along the horizontal axis. The 0–4 age group was represented by a bar of horizontal width 0.5 cm and height 5.2 cm.

(a) Find the widths and heights, in centimetres to 1 decimal place, of the bars representing the following age groups:
   (i) 16–44,   (ii) 65–79.
(b) Taking the mid-point of the last group to be 90 years write down the mid-points of the other age groups and estimate the mean age of the population in this region of the United Kingdom.

[Edexcel]

10 Whig and Penn, solicitors, monitored the time spent on consultations with a random sample of 120 of their clients. The times, to the nearest minute, are summarised in the following table.

Time	Number of clients
10–14	2
15–19	5
20–24	17
25–29	33
30–34	27
35–44	25
45–59	7
60–89	3
90–119	1
Total	120

(a) By calculation, obtain estimates of the median and quartiles of this distribution.
(b) Comment on the skewness of the distribution.
(c) Explain briefly why these data are consistent with the distribution of times you might expect in this situation.
(d) Calculate estimates of the mean and variance of the population of times from which these data were obtained.

The solicitors are undecided whether to use the median and quartiles, or the mean and standard deviation, to summarise these data.

(e) State, giving a reason, which you would recommend them to use.
(f) Given that the least time spent with a client was 12 minutes and the longest time was 116 minutes, draw a box plot to represent these data. Use graph paper and show your scale clearly.

Law and Court, another group of solicitors, monitored the times spent with a random sample of their clients. They found that the least time spent with a client was 20 minutes, the longest time was 40 minutes and the quartiles were 24, 30 and 36 minutes respectively.

(g) Using the same graph paper and the same scale draw a box plot to represent these data.
(h) Compare and contrast the two box plots.

[Edexcel]

KEY POINTS

1. The mean, median and mode or modal class are measures of central tendency.

2. The *mean*, $\bar{x} = \dfrac{\sum x}{n}$. For grouped data $\bar{x} = \dfrac{\sum xf}{n}$.

3. The *median* is the mid-value when the data are presented in rank order; it is the value of the $\frac{n+1}{2}$ th item of n data items.

4. The *mode* is the most common item of data. The *modal class* is the class containing the most data, when the classes are of equal width.

5. The range, the variance and the standard deviation are measures of *spread* or *dispersion*.

6. *Range* = maximum data value − minimum data value.

7. The variance, $sd^2 = \dfrac{1}{n}\sum (x - \bar{x})^2 f$ or $\dfrac{1}{n}\sum x^2 f - \bar{x}^2$.

8. The standard deviation, $sd = \sqrt{\dfrac{1}{n}\sum (x - \bar{x})^2 f}$.

9. If data, represented by the variable x, are coded as $y = a + bx$ then the mean and standard deviation of the coded data are $\bar{y} = a + b\bar{x}$ and $sd_y = b\, sd_x$, respectively.

Chapter four

Probability

If we knew Lady Luck better, Las Vegas would still be a roadstop in the desert.
Stephen Jay Gould

Measuring probability

Probability (or chance) is a way of describing the likelihood of different possible *outcomes* occurring as a result of some *experiment*.

It is important in probability to distinguish experiments from the outcomes which they may generate. Here are a few examples.

Experiments	Possible outcomes
• guessing the answer to a four-option multiple choice question	A B C D
• predicting the stamp on the next letter I receive	first class second class foreign other
• tossing a coin	heads tails

Another word for experiment is *trial*.

Another word you should know is *event*. This often describes several outcomes put together. For example, when rolling a die, an event could be 'the die shows an even number'. This event corresponds to three different outcomes from the trial, the die showing 2, 4 or 6. However, the term event is also often used to describe a single outcome.

EXAMPLE 4.1

Using the notation described above, write down the probability that the correct answer for the next four-option multiple choice question will be answer *A*. What assumptions are you making?

Solution Assuming that the test-setter has used each letter equally often, the probability, P(*A*), that the next question will have answer *A* can be written as follows:

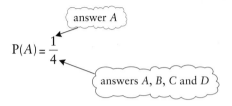

Notice that we have assumed that the four options are equally likely. Equiprobability is an important assumption underlying most work on probability.

Expressed formally, the probability, P(*A*), of event *A* occurring is:

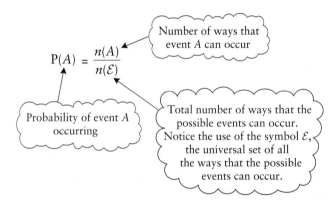

PROBABILITIES OF 0 AND 1

The two extremes of probability are *certainty* at one end of the scale and *impossibility* at the other. Here are examples of certain and impossible events.

Experiments	Certain events	Impossible events
• tossing a single die	the result is in the range 1 to 6 inclusive	the result is a 7
• tossing a coin	getting either heads or tails	getting neither heads nor tails

Certainty

As you can see from the table above, for events that are certain, the number of ways that the event can occur, $n(A)$ in the formula, is equal to the total number of possible events, $n(\mathcal{E})$.

$$\frac{n(A)}{n(\mathcal{E})} = \frac{n(\mathcal{E})}{n(\mathcal{E})} = 1$$

So the probability of an event which is certain is 1.

Impossibility

For impossible events, the number of ways that the event can occur, $n(A)$, is zero.

$$\frac{n(A)}{n(\mathcal{E})} = \frac{0}{n(\mathcal{E})} = 0$$

So the probability of an event which is impossible is 0.

Typical values of probabilities might be something like 0.3 or 0.9. If you arrive at probability values of, say, −0.4 or 1.7, you will know that you have made a mistake since these are meaningless.

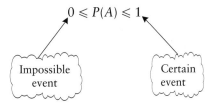

THE COMPLEMENT OF AN EVENT

The complement of an event A, denoted by A', is the event *not-A*, that is the event 'A does not happen'.

EXAMPLE 4.2

It was found that, out of a box of 50 matches, 45 lit but the others did not. What was the probability that a randomly selected match would not have lit?

Solution   The probability that a randomly selected match lit was

$$P(A) = \frac{45}{50} = 0.9.$$

The probability that a randomly selected match did not light was

$$P(A') = \frac{(50-45)}{50} = \frac{5}{50} = 0.1.$$

From this example you can see that

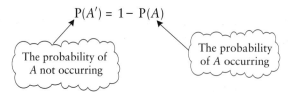

$$P(A') = 1 - P(A)$$

The probability of A not occurring

The probability of A occurring

This is illustrated in figure 4.1.

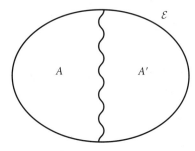

FIGURE 4.1   *Venn diagram showing events A and Not-A(A′)*

EXPECTATION

Suppose that a candidate for a 20-question multiple choice test decides to randomly guess A, B, C or D from the four choices available. How many questions can he expect to get right?

The probability p of guessing any particular question correctly is $\frac{1}{4}$. Since there are 20 questions, a reasonable figure for the expected total number of right answers might be $20 \times \frac{1}{4} = 5$.

This quantity is called the *expectation* or *expected value*. It is found from the product np, where n is the number of trials and p is the probability of an individual success.

Expectation is a technical term and need not be a whole number. Thus the expectation of the number of heads when a coin is tossed 5 times is $5 \times \frac{1}{2} = 2.5$. You would be wrong to go on to say 'That means either 2 or 3' or to qualify your answer as 'about $2\frac{1}{2}$'. The expectation is 2.5.

Expectation is often used in the context of winnings from a gambling game.

EXAMPLE 4.3

In a raffle 500 tickets are sold for £1 each. There are five winning tickets. One ticket wins the first prize of £50, the other four each win a prize of £20. What are the expected winnings from buying a ticket?

Solution

$$\text{Expected winnings} = £(50 \times \tfrac{1}{500} + 20 \times \tfrac{4}{500}) = £0.26$$

Since a ticket costs £1, anybody buying tickets can expect to lose on average 74p for each ticket they buy.

THE PROBABILITY OF EITHER ONE EVENT OR ANOTHER

So far we have looked at just one event at a time. However, it is often useful to bracket two or more of the events together and calculate their combined probability.

EXAMPLE 4.4

A library owns 80 000 books; each book is either on the shelves, out on loan, or on unauthorised loan. The table shows the number of books in each category, with a corresponding probability.

Category of book	Typical numbers	Probability
On the shelves (S)	20 000	0.25
Out on loan (L)	44 000	0.55
Unauthorised loan (U)	16 000	0.20
Total ($S + L + U$)	80 000	1.00

What is the probability that a randomly requested book is *either* out on loan *or* on unauthorised loan (i.e. that it is not available)?

Solution

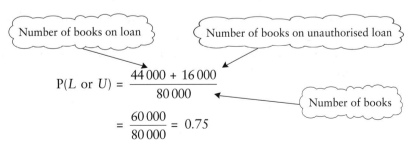

This can be written in more formal notation as

$$P(L \cup U) = \frac{n(L \cup U)}{n(\mathcal{E})} = \frac{n(L)}{n(\mathcal{E})} + \frac{n(U)}{n(\mathcal{E})} = P(L) + P(U)$$

Notice the use of the *union* symbol, ∪, to mean *or*. This is illustrated in figure 4.2.

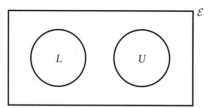

Key: L = out on loan
U = out on unauthorised loan

FIGURE 4.2 *Venn diagram showing events L and U. It is not possible for both to occur*

In this example you could add the probabilities of the two events to get the combined probability of *either one or the other* event occurring. However, you have to be very careful adding probabilities as you will see in the next example.

EXAMPLE 4.5

Below are further details of the categories of books in the library.

Category of book	Number of books
On the shelves	20 000
Out on loan	44 000
Adult fiction	22 000
Adult non-fiction	40 000
Junior	18 000
Unauthorised loan	16 000
Total stock	80 000

Assuming all the books in the library are equally likely to be requested, find the probability that the next book requested will be either out on loan or a book of adult non-fiction.

Solution

$$P(\text{on loan}) + P(\text{adult non-fiction}) = \frac{44\,000}{80\,000} + \frac{40\,000}{80\,000}$$

$$= 0.55 + 0.5 = 1.05$$

This is clearly nonsense as you cannot have a probability greater than 1.

So what has gone wrong?

The way this calculation was carried out involved some double counting. Some of the books classed as adult non-fiction were counted twice because they were also in the on-loan category, as you can see from figure 4.3.

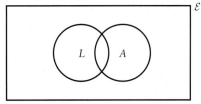

Key: L = out on loan
A = adult non-fiction

FIGURE 4.3 *Venn diagram showing events L and A. It is possible for both to occur*

If you add all six of the book categories together, you find that they add up to 160 000, which represents twice the total number of books owned by the library.

A more useful representation of these data is given in the two-way table below.

	Adult fiction	Adult non-fiction	Junior	Total
On the shelves	4 000	12 000	4 000	20 000
Out on loan	14 000	20 000	10 000	44 000
Unauthorised loan	4 000	8 000	4 000	16 000
Total	22 000	40 000	18 000	80 000

If you simply add 44 000 and 40 000, you *double count* the 20 000 books which fall into both categories. So you need to subtract the 20 000 to ensure that it is counted only once. Thus:

Number either out on loan or adult non-fiction

$$= 44\,000 + 40\,000 - 20\,000$$

$$= 64\,000 \text{ books.}$$

So, the required probability $= \frac{64\,000}{80\,000} = 0.8$.

MUTUALLY EXCLUSIVE EVENTS

The problem of double counting does not occur when adding two rows in the table. Two rows cannot overlap, or *intersect*, which means that those categories are *mutually exclusive* (i.e. the one excludes the other). The same is true for two columns within the table.

Where two events, A and B, are mutually exclusive, the probability that either A or B occurs is equal to the sum of the separate probabilities of A and B occurring.

Where two events, A and B, are *not* mutually exclusive, the probability that either A or B occurs is equal to the sum of the separate probabilities of A and B occurring minus the probability of A and B occurring together.

$P(A \cup B) = P(A) + P(B)$ $P(A \cup B) = P(A) + P(B) - P(A \cap B)$

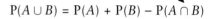

Notice the use of the intersection sign, ∩, to mean *both* ... *and* ...

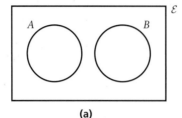
(a)

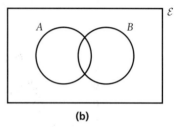
(b)

FIGURE 4.4 **(a)** *Mutually exclusive events;* **(b)** *Not mutually exclusive events*

EXAMPLE 4.6

A fair die is thrown. What is the probability that it shows
(a) Event A: an even number
(b) Event B: a number greater than 4
(c) Event $A \cup B$: a number which is either even or greater than 4?

Solution (a) Event A:
Three out of the six numbers on a die are even, namely 2, 4 and 6.

So $P(A) = \frac{3}{6} = \frac{1}{2}$.

(b) Event B:
Two out of the six numbers on a die are greater than 4, namely 5 and 6.

So $P(B) = \frac{2}{6} = \frac{1}{3}$.

(c) Event $A \cup B$:
Four of the numbers on a die are either even or greater than 4, namely 2, 4, 5 and 6.

So $P(A \cup B) = \frac{4}{6} = \frac{2}{3}$.

This could also be found using

$$P(A \cup B) = P(A) + P(B) - P(A \cap B)$$
$$P(A \cup B) = \frac{3}{6} + \frac{2}{6} - \frac{1}{6}$$
$$= \frac{4}{6} = \frac{2}{3}$$

This is the number 6 which is both even and greater than 4.

Exercise 4A

1. Three separate electrical components, switch, bulb and contact point, are used together in the construction of a pocket torch. Of 534 defective torches, examined to identify the cause of failure, 468 are found to have a defective bulb. For a given failure of the torch, what is the probability that either the switch or the contact point is responsible for the failure? State clearly any assumptions that you have made in making this calculation.

2. If a fair die is thrown, what is the probability that it shows
   (a) 4
   (b) 4 or more
   (c) less than 4
   (d) an even number?

3. A bag containing Scrabble letters has the following letter distribution:

A	B	C	D	E	F	G	H	I	J	K	L	M
9	2	2	4	12	2	3	2	9	1	1	4	2

N	O	P	Q	R	S	T	U	V	W	X	Y	Z
6	8	2	1	6	4	6	4	2	2	1	2	1

   The first letter is chosen at random from the bag; find the probability that it is
   (a) an E
   (b) in the first half of the alphabet
   (c) in the second half of the alphabet
   (d) a vowel
   (e) a consonant
   (f) the only one of its kind.

4. A lottery offers five prizes, each of £100, and a total of 2000 lottery tickets are sold. You buy a single ticket for 20p.
   (a) What is the probability that you will win a prize?
   (b) What is the probability that you will not win a prize?
   (c) How much money do the lottery organisers expect to make or lose?
   (d) How much money should the lottery organisers charge for a single ticket in order to break even?
   (e) If they continue to charge 20p per ticket, how many tickets would they need to sell in order to break even?

5. **A sporting chance**
   (a) Two players, A and B, play tennis. On the basis of their previous results, the probability of A winning, P(A), is calculated to be 0.65. What is P(B), the probability of B winning?
   (b) Two hockey teams, A and B, play a game. On the basis of their previous results, the probability of team A winning, P(A), is calculated to be 0.65. Why is it not possible to calculate directly P(B), the probability of team B winning, without further information?

(c) In a tennis tournament, player A, the favourite, is estimated to have a 0.3 chance of winning the competition. Player B is estimated to have a 0.15 chance. Find the probability that either A or B will win the competition.

(d) In the Six Nations Rugby Championship, France and England are given a 25% chance of winning or sharing the championship cup. It is also estimated that there is a 5% chance that they will share the cup. Estimate the probability that either England or France will win or share the cup.

6 The integers 1 to 20 are classified as being either even (E), odd (O) or square (S). (Some numbers are in more than one category.) They are written on separate identical cards and the cards are then thoroughly shuffled.

(a) Represent on a Venn diagram the possible outcomes of drawing a card at random.

(b) A card is chosen at random. Find the probability that the number showing is:
   (i)   even, E (ii) square, S (iii) odd, O
   (iv) both even and square, $E \cap S$
   (v)   either even or square, $E \cup S$
   (vi) both even and odd, $E \cap O$
   (vii) either even or odd, $E \cup O$.

   Write down equations connecting the probabilities of the following events:
   (viii) $E, S, E \cap S, E \cup S$
   (ix) $E, O, E \cap O, E \cup O$.

7 The data in the table below show the numbers of part-time students in higher education in 1988/89 by sex and type of establishment (numbers in thousands).

	Women	Men
Universities	21.1	29.0
Open University	40.3	45.0
Polytechnics and colleges		
– part-time day courses	67.0	118.5
– evening only courses	26.5	38.1
Total part-time students	154.9	230.6

(Source: *Education Statistics for the United Kingdom*, Department of Education and Science.)

Find the probability that a part-time student chosen at random is
(a) an Open University student
(b) female
(c) a female Open University student
(d) studying at a polytechnic or college
(e) not studying at a polytechnic or college.

The probability of events from two trials

To calculate the probability of events from two trials a tree diagram is often used.

Probabilities are multiplied along the branches to produce the final figure at the right hand end of each path.

EXAMPLE 4.7

Veronica attends a summer fair. She buys 1 ticket of the 1245 raffle tickets sold, and 1 programme of the 324 programmes sold. Later that day two prize draws are held, one to select the winner of the raffle and the other to select the winner of the programme draw.

Find the probability that
(a) Veronica wins both draws
(b) Veronica wins neither draw
(c) Veronica wins one of the two draws.

Solution The possible results are shown on the tree diagram in figure 4.5.

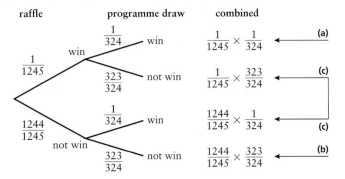

FIGURE 4.5

(a) The probability that Veronica wins both

$$= \frac{1}{1245} \times \frac{1}{324} = \frac{1}{403\,380}$$

This is not quite 'one in a million' but is not very far off it.

(b) The probability that Veronica wins neither

$$= \frac{1244}{1245} \times \frac{323}{324} = \frac{401\,812}{403\,380}$$

This of course is much the most likely outcome.

(c) The probability that Veronica wins one but not the other is given by

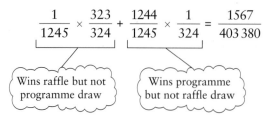

Look again at the structure of the tree diagram in figure 4.5.

There are two experiments, the raffle draw and the programme draw. These are considered as *First, Then* experiments, and set out *First* on the left and *Then* on the right. Once you understand this, the rest of the layout falls into place, with the different outcomes or events appearing as branches. In this example there are two branches at each stage; sometimes there may be three or more. Notice that for a given situation the component probabilities sum to 1, as before.

$$\frac{1}{403\,380} + \frac{323}{403\,380} + \frac{1244}{403\,380} + \frac{401\,812}{403\,380} = \frac{403\,380}{403\,380} = 1$$

EXAMPLE 4.8

Some friends buy a six-pack of potato crisps. Two of the bags are snake flavoured (S), the rest are frog flavoured (F). They decide to allocate the bags by lucky dip. Find the probability that

(a) the first two bags chosen are the same as each other
(b) the first two bags chosen are different from each other.

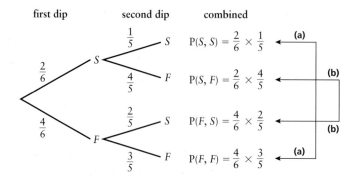

FIGURE 4.6

The probability of events from two trials

Solution Note: $P(F, S)$ means the probability of drawing a frog bag (F) on the first dip and a snake bag (S) on the second.

(a) The probability that the first two bags chosen are the same as each other = $P(S, S) + P(F, F) = \frac{2}{6} \times \frac{1}{5} + \frac{4}{6} \times \frac{3}{5}$

$= \frac{1}{15} + \frac{6}{15}$

$= \frac{7}{15}$

(b) The probability that the first two bags chosen are different from each other = $P(S, F) + P(F, S) = \frac{2}{6} \times \frac{4}{5} + \frac{4}{6} \times \frac{2}{5}$

$= \frac{4}{15} + \frac{4}{15}$

$= \frac{8}{15}$

Note The answer to part (b) above hinged on the fact that two orderings (S then F, and F then S) are possible for the same combined event (that the two bags selected include one snake and one frog bag).

The probabilities changed between the first dip and the second dip. This is because the outcome of the second dip is *dependent* on the outcome of the first one (with fewer bags remaining to choose from).

By contrast, the outcomes of the two experiments involved in tossing a coin twice are *independent*, and so the probability of getting a head on the second toss remains unchanged at 0.5, whatever the outcome of the first toss.

Although you may find it helpful to think about combined events in terms of how they would be represented on a tree diagram, you may not always actually draw them in this way. If there are several experiments and perhaps more than two possible outcomes from each, drawing a tree diagram can be very time-consuming.

Tree diagrams are not the only approach to questions involving the probability of events from two trials. Sometimes it is appropriate to draw a *sample space diagram*, as in this next example.

EXAMPLE 4.9

A six-sided die is numbered so that its faces show the numbers 1, 2, 2, 3, 3, 6. It is thrown twice. Find the probability that the total score is

(a) 5

(b) even

(c) at least 6.

Solution

(a) There are 36 equally likely combinations of scores from the two throws:

	1	2	2	3	3	6
1	2	3	3	4	4	7
2	3	4	4	5	5	8
2	3	4	4	5	5	8
3	4	5	5	6	6	9
3	4	5	5	6	6	9
6	7	8	8	9	9	12

This type of grid is called a sample space diagram

FIGURE 4.7

$$\therefore \quad P(\text{total} = 5) = \frac{8}{36} = \frac{2}{9}$$

(b) 18 of the 36 totals are even.

$$\therefore \quad P(\text{total is even}) = \frac{18}{36} = \frac{1}{2}$$

(c)

	1	2	2	3	3	6
1	2	3	3	4	4	7
2	3	4	4	5	5	8
2	3	4	4	5	5	8
3	4	5	5	6	6	9
3	4	5	5	6	6	9
6	7	8	8	9	9	12

This region contains totals of at least 6; there are 15 of them

FIGURE 4.8

$$\therefore \quad P(\text{total is at least } 6) = \frac{15}{36} = \frac{5}{12}$$

EXERCISE 4B

1 The probability of a pregnant woman giving birth to a baby girl is about 0.49. Draw a tree diagram showing the possible outcomes if she has two babies (not twins). From the tree diagram, calculate the following probabilities:
   (a) that the babies are both girls
   (b) that the babies are the same sex
   (c) that the second baby is of different sex to the first.

2 In a certain district of a large city, the probability of a household suffering a break-in in a particular year is 0.07 and the probability of its car being stolen is 0.12. Assuming these two trials are independent of each other, draw a tree diagram showing the possible outcomes for a particular year. Calculate, for a randomly selected household with one car, the following probabilities:
   (a) that the household is a victim of both crimes during that year
   (b) that the household suffers *only one* of these misfortunes during that year
   (c) that the household suffers *at least one* of these misfortunes during that year.

EXERCISE 4B

3 There are 12 people at an identification parade. Three witnesses are called to identify the accused person. Assuming they make their choice purely by random selection, draw a tree diagram showing the possible events.
  (a) From the tree diagram, calculate the following probabilities:
    (i) that all three witnesses select the accused person
    (ii) that none of the witnesses selects the accused person
    (iii) that at least two of the witnesses select the accused person.
  (b) Suppose now that by changing the composition of people in the identification parade, the first two witnesses increase their chances of selecting the accused person to 0.25. Draw a new tree diagram and calculate the following probabilities:
    (i) that all three witnesses select the accused person
    (ii) that none of the witnesses selects the accused person
    (iii) that at least two of the witnesses select the accused person.

4 Ruth drives her car to work – provided she can get it to start! When she remembers to put the car in the garage the night before, it starts next morning with a probability of 0.95. When she forgets to put the car away, it starts next morning with a probability of 0.75. She remembers to garage her car 90% of the time.

What is the probability that Ruth drives her car to work on a randomly chosen day?

5 Around 0.8% of men are red–green colour-blind (the figure is slightly different for women) and roughly 1 in 5 men is left-handed. Assuming these characteristics are inherited independently, calculate with the aid of a tree diagram the probability that a man chosen at random will
  (a) be both colour-blind and left-handed
  (b) be colour-blind and not left-handed
  (c) be colour-blind or left-handed
  (d) be neither colour-blind nor left-handed.

6 A gambling game consists of tossing a coin three times. You win if all three tosses give the same result (i.e. three heads or three tails) and you lose if any other event shows. Calculate the probability that you will lose a particular game.

7 All the Jacks, Queens and Kings are removed from a pack of cards. Giving the Ace a value of 1, this leaves a pack of 40 cards consisting of four suits of cards numbered 1 to 10. The cards are well shuffled and one is drawn and noted. This card is not returned to the pack and a second card is drawn. Find the probability that
  (a) both cards are even
  (b) at least one card is odd
  (c) both cards are of the same suit
  (d) only one of the cards has a value greater than 7.

8 Three dice are thrown. Find the probability of obtaining
  (a) at least two 6s
  (b) no 6s
  (c) different scores on all the dice.

9 Explain the flaw in this argument and rewrite it as a valid statement.
The probability of throwing a 6 on a fair die $= \frac{1}{6}$. Therefore the probability of throwing at least one 6 in six throws of the die is
$\frac{1}{6} + \frac{1}{6} + \frac{1}{6} + \frac{1}{6} + \frac{1}{6} + \frac{1}{6} = 1$ *so it is a certainty.*

10 In a Donkey Derby event, there are three races. There are six donkeys entered for the first race, four for the second and three for the third. Sheila places a bet on one donkey in each race. She knows nothing about donkeys and chooses each donkey at random. Find the probability that she backs at least one winner.

11 Two dice are thrown, one red and the other green.
  (a) Copy and complete this table showing all the possible outcomes.

		\multicolumn{6}{c}{Green die}
	+	1	2	3	4	5	6
Red die	1						
	2						
	3						
	4						10
	5						11
	6	7	8	9	10	11	12

  (b) What is the probability of a score of 4?
  (c) What is the most likely outcome?
  (d) Criticise this argument:
    There are 11 possible outcomes, 2, 3, 4, up to 12. Therefore each of them has a probability of $\frac{1}{11}$.

12 A gambling game consists of tossing a coin three times. If two or more heads show, you get your money back. If you throw three heads, you receive double your stake money. Otherwise you lose. For a single throw, calculate
  (a) the probability of losing your money
  (b) the probability of doubling your money
  (c) the probability of just getting your money back
  (d) the expectation of the value of your winnings, given a £1 stake.

13 The probability of someone catching flu in a particular winter when they have been given the flu vaccine is 0.1. Without the vaccine, the probability of catching flu is 0.4. If 30% of the population has been given the vaccine, what is the probability that a person chosen at random from the population will catch flu over that winter?

Conditional probability

You have already used the multiplication principle for independent events:

$$P(A \cap B) = P(A) \times P(B)$$

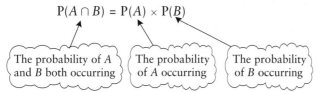

The probability of A and B both occurring / The probability of A occurring / The probability of B occurring

If the events A and B are not independent then the result must be modified:

$$P(A \cap B) = P(A) \times P(B \mid A)$$

The probability of B given A

This result may be rewritten as

$$P(B \mid A) = \frac{P(A \cap B)}{P(A)}$$

In words, you say:

$$\text{'probability of } B \text{ given } A\text{'} = \frac{\text{'probability of } A \text{ and } B\text{'}}{\text{'probability of } A\text{'}}$$

EXAMPLE 4.10

A company is worried about the high turnover of its employees and decides to investigate whether they are more likely to stay if they are given training.
On 1 January one year the company employed 256 people (excluding those about to retire). During that year a record was kept of who received training as well as who left the company. The results are summarised in this table:

	Still employed	Left company	
Given training	109	43	152
Not given training	60	44	104
	169	87	256

Find the probability that a randomly selected employee
(a) received training
(b) did not leave the company
(c) received training and did not leave the company
(d) did not leave the company, given that the person had received training
(e) did not leave the company, given that the person had not received training.

Solution Using the notation T: The employee received training
S: The employee stayed in the company

(a) $P(T) = \dfrac{n(T)}{n(\mathcal{E})} = \dfrac{152}{256}$

(b) $P(S) = \dfrac{n(S)}{n(\mathcal{E})} = \dfrac{169}{256}$

(c) $P(T \cap S) = \dfrac{n(T \cap S)}{n(\mathcal{E})} = \dfrac{109}{256}$

(d) $P(S \mid T) = \dfrac{P(S \cap T)}{P(T)} = \dfrac{\frac{109}{256}}{\frac{152}{256}} = \dfrac{109}{152} = 0.72$

(e) $P(S \mid T') = \dfrac{P(S \cap T')}{P(T')} = \dfrac{\frac{60}{256}}{\frac{104}{256}} = \dfrac{60}{104} = 0.58$

Since $P(S \mid T)$ is not the same as $P(S \mid T')$, the event S is not independent of the event T. Each of S and T is dependent on the other, a conclusion which matches common sense. It is almost certainly true that training increases employees' job satisfaction and so makes them more likely to stay, but it is also probably true that the company is more likely to go to the expense of training the employees who seem less inclined to move on to other jobs.

In some situations you may find it helps to represent a problem such as this as a Venn diagram.

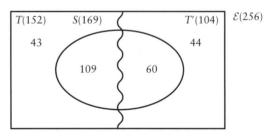

FIGURE 4.9

In other situations it may be helpful to think of conditional probabilities in terms of tree diagrams. Conditional probabilities are needed when events are *dependent*, that is when the outcome of one trial affects the outcomes from a subsequent trial, so, for dependent events, the probabilities of all but the first layer of a tree diagram will be conditional.

EXAMPLE 4.11

Rebecca is buying two goldfish from a pet shop. The shop's tank contains seven male fish and eight female fish but they all look the same.

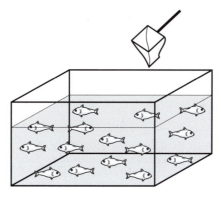

FIGURE 4.10

Find the probability that Rebecca's fish are
(a) both the same sex
(b) both female
(c) both female given that they are the same sex.

Solution The situation is shown on this tree diagram.

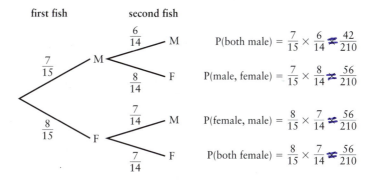

FIGURE 4.11

(a) P(both the same sex) = P(both male) + P(both female)

$$= \frac{42}{210} + \frac{56}{210} = \frac{98}{210} = \frac{7}{15}$$

(b) P(both female) = $\frac{56}{210} = \frac{4}{15}$

(c) P(both female | both the same sex)

= P(both female and the same sex)/P(both the same sex) = $\dfrac{\frac{4}{15}}{\frac{7}{15}} = \frac{4}{7}$

(This is the same as P(both female))

The ideas in the last example can be expressed more generally for any two dependent events, A and B. The tree diagram would be as shown in figure 4.10.

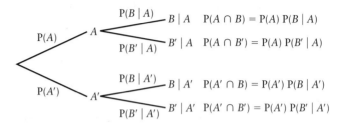

FIGURE 4.12

{The probabilities in the second layer of the tree diagram are conditional on the outcome of the first experiment}

{These events are conditional upon the outcome of the first experiment}

The tree diagram shows you that
- $P(B) = P(A \cap B) + P(A' \cap B)$
  $= P(A)P(B \mid A) + P(A')P(B \mid A')$
- $P(A \cap B) = P(A)P(B \mid A)$

$\Rightarrow P(B \mid A) = \dfrac{P(A \cap B)}{P(A)}$

EXERCISE 4C

1 In a school of 600 pupils, 360 are girls. There are 320 hockey players, of whom 200 are girls. Among the hockey players there are 28 goalkeepers, 19 of them girls. Find the probability that
  (a) a pupil chosen at random is a girl
  (b) a girl chosen at random plays hockey
  (c) a hockey player chosen at random is a girl
  (d) a pupil chosen at random is a goalkeeper
  (e) a goalkeeper chosen at random is a boy
  (f) a male hockey player chosen at random is a goalkeeper
  (g) a hockey player chosen at random is a male goalkeeper
  (h) two pupils chosen at random are both goalkeepers
  (i) two pupils chosen at random are a male goalkeeper and a female goalkeeper
  (j) two pupils chosen at random are one boy and one girl.

2 The table gives the numbers of offenders in England and Wales sentenced for indictable offences, by type of offence and type of sentence, in 1989 (numbers in thousands).

	Discharge	Probation	Fine	Jail	Other	Total
Violence	9.5	3.9	20.6	9.0	12.7	55.7
Sexual offences	0.8	0.9	2.4	2.4	0.8	7.3
Theft	31.4	26.2	62.6	29.9	32.1	182.2
Motoring	0.5	0.3	8.7	1.0	0.8	11.3
Other	12.0	9.4	41.3	11.9	7.9	82.5
Total	54.2	40.7	135.6	54.2	54.3	339.0

(Source: adapted from Table 12.19, *Social Trends* 22, 1992, CSO.)

(a) Find the probability that a randomly selected person indicted for an offence will be
   (i) discharged   (ii) put on probation
   (iii) fined   (iv) sent to jail
   (v) sent to jail for a motoring offence
   (vi) sent to jail given that the person has committed a motoring offence
   (vii) guilty of a motoring offence given that the person is sent to jail.

(b) Criticise this statement:
   Based on these figures nearly 2% of the country's prison population are there for motoring offences.

3 100 cars are entered for a road-worthiness test which is in two parts, mechanical and electrical. A car passes only if it passes both parts. Half the cars fail the electrical test and 62 pass the mechanical; 15 pass the electrical but fail the mechanical test. Find the probability that a car chosen at random
   (a) passes overall
   (b) fails on one test only
   (c) given that it has failed, failed the mechanical test only.

4 Two dice are thrown. What is the probability that the total is
   (a) 7
   (b) a prime number
   (c) 7, given that it is a prime number?

5 A cage holds two litters of rats. One litter comprises three females and four males, and the other comprises two females and six males. A random selection of two rats is made. Find, as fractions, the probabilities that the two rats are
   (a) from the same litter
   (b) of the same sex
   (c) from the same litter and of the same sex
   (d) from the same litter given that they are of the same sex.

[MEI]

6 In a school of 400 pupils, 250 play a musical instrument and 100 sing in the choir.

The probability that a pupil chosen at random neither plays a musical instrument nor sings in the choir is $\frac{1}{5}$.

(a) How many pupils both sing in the choir and play a musical instrument?
(b) Find the probability that a pupil chosen at random sings in the choir but does not play an instrument.
(c) Find the probability that a member of the choir chosen at random does not play an instrument.
(d) Find the probability that someone who does not play an instrument, chosen at random, is in the choir.

7 A bag P contains three red balls. A second bag Q contains two red balls and three black balls.

(a) A bag is chosen at random and one ball is withdrawn. Find the probability that this ball is red.

This ball remains outside the bag.

(b) A bag is again chosen at random (it is not known whether this is the same bag as before or not) and one ball is withdrawn. Find the joint probability that both this ball and the one previously withdrawn are red.
(c) If they are both red, what is the probability that bag P was used on both occasions?

[O & C]

8 In a child's game there should be seven triangles, three of which are blue and four of which are red, and eleven squares, five of which are blue and six of which are red. However, two pieces are lost. Assuming the pieces are lost at random, find the probability that they are

(a) the same shape
(b) the same colour
(c) the same shape and the same colour
(d) the same shape given that they are the same colour.

[MEI]

9 A and B are two events with probabilities given by $P(A) = 0.4$, $P(B) = 0.7$ and $P(A \cap B) = 0.35$.

(a) Find $P(A \mid B)$ and $P(B \mid A)$.
(b) Show that the events A and B are not independent.

10 Quark hunting is a dangerous occupation. On a quark hunt, there is a probability of $\frac{1}{4}$ that the hunter is killed. The quark is twice as likely to be killed as the hunter. There is a probability of $\frac{1}{3}$ that both survive.

(a) Copy and complete this table of probabilities.

	Hunter dies	Hunter lives	
Quark dies			$\frac{1}{2}$
Quark lives		$\frac{1}{3}$	$\frac{1}{2}$
	$\frac{1}{4}$		1

Find the probability that
(b) both the hunter and the quark die
(c) the hunter lives and the quark dies
(d) the hunter lives, given that the quark dies.

EXERCISE 4D   **Examination style questions**

1 A contractor bids for two projects. He estimates that the probability of winning the first project is 0.5, the probability of winning the second project is 0.3 and the probability of winning both projects is 0.2.
(a) Find the probability that he does not win either project.
(b) Find the probability that he wins exactly one project.
(c) Given that he does not win the first project, find the probability that he wins the second.
(d) By calculation, determine whether or not winning the first contract and winning the second contract are independent events.

[Edexcel]

2 Every year two teams, the *Ramblers* and the *Strollers*, meet each other for a quiz night. From past results it seems that in years when the *Ramblers* win, the probability of them winning the next year is 0.7 and in years when the *Strollers* win, the probability of them winning the next year is 0.5. It is not possible for the quiz to result in the scores being tied.

The *Ramblers* won the quiz in 1996.
(a) Draw a probability tree diagram for the three years up to 1999.
(b) Find the probability that the *Strollers* will win in 1999.
(c) If the *Strollers* win in 1999, what is the probability that it will be their first win for at least three years?
(d) Assuming that the *Strollers* win in 1999, find the smallest value of n such that the probability of the *Ramblers* winning the quiz for n consecutive years after 1999 is less than 5%.

[MEI]

3 On my way home from work each evening I have to pass through three sets of traffic lights in the city centre. The probabilities that I pass through them *without* having to stop are 0.2, 0.4 and 0.7 respectively. You may assume that each set of lights operates independently of the others.
   (a) Draw a tree diagram to illustrate the situation.
   (b) Find the probability that I do not have to stop at any of the three sets of lights.
   (c) Find the probability that I have to stop at just one set of lights.
   (d) Given that I have to stop at just one set of lights, find the probability that I have to stop at the first set of lights.
   (e) It is decided to change the probability of passing through the first lights without having to stop from 0.2 to p. I find that this change reduces the probability found in part (d) to 0.5. Calculate p.

[MEI]

4 There are 90 players in a tennis club. Of these, 23 are juniors, the rest are seniors. 34 of the seniors and 10 of the juniors are male. There are 8 juniors who are left-handed, 5 of whom are male. There are 18 left-handed players in total, 4 of whom are female seniors.
   (a) Represent this information in a Venn diagram.
   (b) What is the probability that:
      (i) a male player selected at random is left-handed?
      (ii) a left-handed player selected at random is a female junior?
      (iii) a player selected at random is either a junior or a female?
      (iv) a player selected at random is right-handed?
      (v) a right-handed player selected at random is not a junior?
      (vi) a right-handed female player selected at random is a junior?

5 The three events E_1, E_2 and E_3 are defined in the same sample space. The events E_1 and E_3 are mutually exclusive. The events E_1 and E_2 are independent.

Given that $P(E_1) = \frac{2}{5}$, $P(E_3) = \frac{1}{3}$ and $P(E_1 \cup E_2) = \frac{5}{8}$, find
   (a) $P(E_1 \cup E_3)$,
   (b) $P(E_2)$.

[Edexcel]

6 An urn A contains 5 red balls and 3 green balls and a similar urn B contains 3 red balls and 5 green balls. One ball is selected at random from A and placed in B. One ball is then selected at random from B. Let R_i and G_i, $i = 1, 2$, represent the events that the ith ball selected is red and green respectively.
   (a) Show that $P(R_1 \cap R_2) = \frac{5}{18}$.
   (b) Find the probability that the second ball selected is red.
   (c) Calculate $P(G_1|G_2)$.

[Edexcel]

Exercise 4D

7. In a large garden there are 7 fruit trees and 13 trees of other types. Six of the trees have birds nesting in them but only two of these are fruit trees.

   (a) Copy and complete the table below to illustrate this information.

	Fruit tree	Other tree	Total
Bird's nest	2		6
No nest			
Total	7	13	

   The owner of the garden has given permission for Abdul to play in the garden but has instructed him not to climb any fruit trees that have birds nesting in them. Abdul selects a tree at random to climb.

   (b) Find the probability that Abdul will obey the owner's instructions.

   Given that Abdul climbs a fruit tree,
   (c) find the probability that the tree has birds nesting in it.

   [Edexcel]

8. The probability that for any married couple the husband has a degree is $\frac{6}{10}$ and the probability that the wife has a degree is $\frac{1}{2}$. The probability that the husband has a degree, given that the wife has a degree, is $\frac{11}{12}$.

   A married couple is chosen at random.

   Find the probability that
   (a) both of them have degrees
   (b) only one of them has a degree
   (c) neither of them has a degree.

   Two married couples are chosen at random.
   (d) Find the probability that only one of the two husbands and only one of the two wives have a degree.

   [Edexcel]

9. The employees of a company are classified as management, administration or production. The following table shows the number employed in each category and whether or not they live close to the company or some distance away.

	Live close	Live some distance away
Management	6	14
Adminstration	25	10
Production	45	25

An employee is chosen at random. Find the probability that this employee
(a) is an administrator
(b) lives close to the company, given that the employee is a manager.

Of the managers, 90% are married, as are 60% of the administrators and 80% of the production employees.
(c) Construct a tree diagram containing all the probabilities.
(d) Find the probability that an employee chosen at random is married.

An employee is selected at random and found to be married.
(e) Find the probability that this employee is in production.

[Edexcel]

10 In a set of 28 dominoes each domino has from 0 to 6 spots at each end. Each domino is different from every other and the ends are indistinguishable so that, for example, the two diagrams below represent the *same* domino.

A domino which has the same number of spots at each end, or no spots at all, is called a 'double'. A domino is drawn at random from the set. The diagram below (left) shows a sample space diagram to represent the complete set of outcomes, each of which is equally likely.

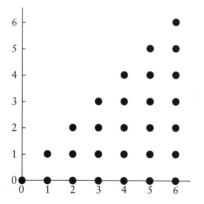 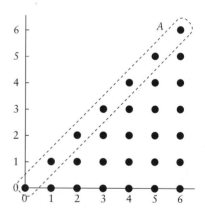

Let the event A be 'the domino is a double', event B be 'the total number of spots on the domino is 6' and event C be 'at least one end of the domino has 5 spots'.

The diagram above (right) shows the sample space with the event A marked.

EXERCISE 4D

(a) Write down the probability that event A occurs.
(b) Find the probability that either B or C or both occur.
(c) Determine whether or not events A and B are independent.
(d) Find the conditional probability $P(A | C)$. Explain why events A and C are *not* independent.

After the first domino has been drawn, a second domino is chosen at random from the remainder.

(e) Find the probability that at least one end of the first domino has the same number of spots as at least one end of the second domino.
[*Hint*: Consider separately the cases where the first domino is a double and where it is not.]

[MEI]

KEY POINTS

1 The probability of an event A, $P(A) = n(A)/n(\mathcal{E})$, where $n(A)$ is the number of ways that A can occur and $n(\mathcal{E})$ is the total number of ways that all possible events can occur, all of which are equally likely.

2 For any two events, A and B, of the same experiment,

$$P(A \cup B) = P(A) + P(B) - P(A \cap B).$$

Where the events are *mutually exclusive* (i.e. where the events do not overlap) the rule still holds but, since $P(A \cap B)$ is now equal to zero, the equation simplifies to:

$$P(A \cap B) = P(A) + P(B).$$

3 Where an experiment produces two or more mutually exclusive events, the probabilities of the separate events sum to 1.

4 $P(A) + P(A') = 1$

5 $P(B | A)$ means the probability of event B occurring given that event A has already occurred.

$$P(B | A) = \frac{P(A \cap B)}{P(A)}$$

6 The probability that event A and then event B occur, in that order, is $P(A) \times P(B | A)$.

7 If event B is independent of event A, $P(B | A) = P(B | A') = P(B)$.

Chapter five

CORRELATION AND REGRESSION

It is now proved beyond doubt that smoking is one of the leading causes of statistics.

John Peers

The data in the following example are a set of pairs of values for two variables, the goals scored and the points totals of all the teams in the 1998–99 football Premiership. This is an example of *bivariate data*, where each item in the population requires the values of two variables. The graph in figure 5.1 is called a *scatter diagram* and this is a common way of showing bivariate data.

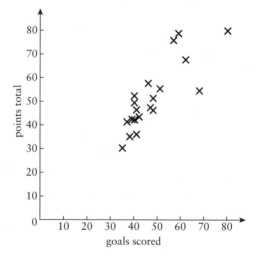

FIGURE 5.1 *Scatter diagram showing goals scored and points total for all teams in the Premiership 1998–99 season*

If each point lies on a straight line, then there is said to be perfect *linear correlation* between the two variables. It is much more likely, however, that your data fall close to a straight line but not exactly on it. The better the fit, the higher the level of linear correlation.

The term *line of best fit* is used to describe a straight line drawn through a set of data points so as to fit them as closely as possible. There are several ways of determining such a line, according to what precisely is meant by a close fit.

Describing Variables

Dependent and independent variables

The scatter diagram in figure 5.1 was drawn with the goals scored on the horizontal axis and the points total on the vertical axis. It was done that way to emphasise that the number of points is dependent upon the number of goals scored. (A team gains points as a result of scoring goals. It does not score goals as a result of gaining points.) It is normal practice to plot the *dependent variable* on the vertical axis and the *independent variable* on the horizontal axis.

The independent variable is sometimes called the *explanatory variable*. The dependent variable is sometimes called the *response variable*.

Here are some more examples of dependent (response) and independent (explanatory) variables.

Independent (explanatory) variable	Dependent (response) variable
Number of people in a lift	Total weight of passengers
The amount of rain falling on a field whilst a crop is growing	The weight of the crop yielded
The number of people visiting a bar in an evening	The volume of beer sold

Random and non-random variables

In the examples above both the variables have unpredictable values and so are *random*.

The same is true for the example about goals scored and points totals in football. Both variables are random variables, free to assume any of a particular set of discrete values in a given range.

Sometimes one or both of the variables is *controlled* so that the variable only assumes a set of predetermined values. Controlled variables are non-random. Here is an example:

Age of a rat in months	6	12	18	24	30
Memory quotient (on a particular scale)	16	22	25	25	23

In this example the age of the rat is a controlled variable. The tests were conducted every six months and not at random times so the age of the rat is a *non-random* variable.

Later in this chapter you will be asked to consider the nature of the variables you are dealing with in more detail.

INTERPRETING SCATTER DIAGRAMS

You can often judge if correlation is present just by looking at a scatter diagram.

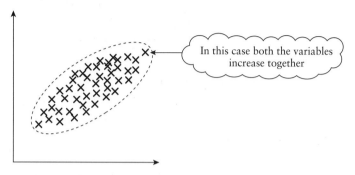

FIGURE 5.2 *Positive correlation*

Notice that in figure 5.2 almost all the observation points can be contained within an ellipse. This shape often arises when both variables are random. You should look for it before going on to do a calculation of Pearson's product moment correlation coefficient (see page 111). The narrower the elliptical profile, the greater is the correlation.

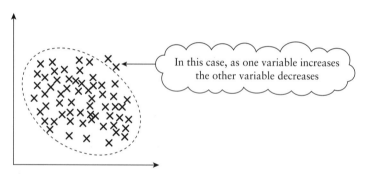

FIGURE 5.3 *Negative correlation*

In figure 5.3 the points again fall into an elliptical profile and this time there is negative correlation. The fatter ellipse in this diagram indicates weaker correlation than in the case shown in figure 5.2.

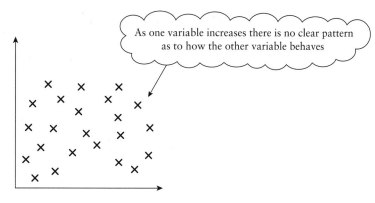

FIGURE 5.4   *No correlation*

In the case illustrated in figure 5.4 the points fall randomly in the (x, y) plane and there appears to be no association between the variables.

You should be aware of some distributions which at first sight appear to indicate linear correlation but in fact do not.

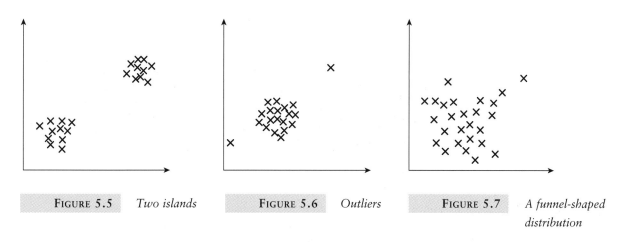

FIGURE 5.5   *Two islands*   FIGURE 5.6   *Outliers*   FIGURE 5.7   *A funnel-shaped distribution*

Figure 5.5 is probably showing two quite different groups, neither of them having any correlation.

Figure 5.6 shows a small data set with no correlation. However, the two outliers give the impression that there is positive linear correlation.

The bulk of the data in Figure 5.7 have no correlation but a few items give the impression that there is correlation.

Note   In none of these three false cases is the distribution even approximately elliptical.

LINE OF BEST FIT

If your scatter diagram leads you to suspect that there is linear correlation between the two variables plotted then you may reasonably try to draw a line of best fit. A simple, but not very accurate, way to do this is as follows.

(a) Calculate and plot the point $(\bar{x}, \bar{y})$ where $\bar{x}$ is the mean of the horizontal axis variable and $\bar{y}$ is the mean of the vertical axis variable.

(b) Draw a straight line which passes through $(\bar{x}, \bar{y})$ and which roughly leaves the same number of points of the scatter diagram above and below it, as in figure 5.8.

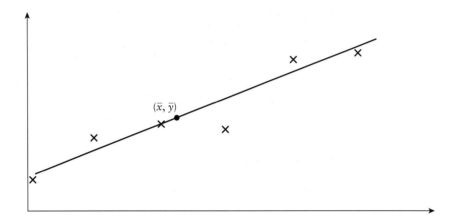

FIGURE 5.8

In some situations this rough and ready method will be adequate, particularly where it is easy to see where the line should go. In others cases, you may find it hard to judge where to place the line and you may well want to draw the line accurately anyway. The method for doing this is described later in this chapter on pages 118–120.

EXERCISE 5A

For each of the sets of data below, draw a scatter diagram and comment on whether there appears to be any correlation. If there is then draw a possible line of best fit.

1 The mathematics and physics test results of 14 pupils.

Mathematics	45	23	78	91	46	27	41	62	34	17	77	49	55	71
Physics	62	36	92	70	67	39	61	40	55	33	65	59	35	40

2 The wine consumption in a country in millions of litres and the years 1993 to 2000.

Year	1993	1994	1995	1996	1997	1998	1999	2000
Consumption ($\times 10^6$ litres)	35.5	37.7	41.5	46.4	44.8	45.8	53.9	62.0

EXERCISE 5A

3 The number of hours of sunshine and the monthly rainfall, in centimetres, in an eight-month period.

	Jan	Feb	Mar	Apr	May	Jun	Jul	Aug
Sunshine (hours)	90	96	105	110	113	120	131	124
Rainfall (cm)	5.1	4.6	6.3	5.1	3.3	2.8	4.5	4.0

4 The annual salary, in thousands of pounds, and the average number of hours worked per week by seven people chosen at random.

Salary (×£1000)	5	7	13	14	16	20	48
Hours worked per week	18	22	35	38	36	36	32

5 The mean temperature in degrees centigrade and the amount of ice-cream sold in a supermarket in hundreds of litres.

	Apr	May	Jun	Jul	Aug	Sep	Oct	Nov
Mean temperature (°C)	9	13	14	17	16	15	13	11
Ice-cream sold (100 litres)	11	15	17	20	22	17	8	7

6 The reaction times of eight women of various ages.

Reaction time ($\times 10^{-3}$ s)	156	165	149	180	189	207	208	178
Age (years)	36	40	27	50	49	53	55	27

PRODUCT MOMENT CORRELATION

The scatter diagram in figure 5.1 revealed that there may be a mutual association between the goals scored and the points totals of teams in the Premiership at the end of the 1998–99 season. In this section we set out to quantify this relationship.

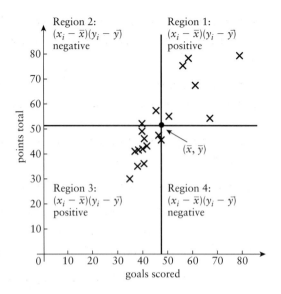

FIGURE 5.9

Figure 5.9 shows the scatter diagram of the last section again.

The mean number of goals scored is

$$\bar{x} = \frac{\sum_{i=1}^{n} x_i}{n} = \frac{80 + 59 + 57 + \cdots}{20} = \frac{959}{20} = 47.95$$

The mean points total is

$$\bar{y} = \frac{\sum_{i=1}^{n} y_i}{n} = \frac{79 + 78 + 75 + \cdots}{20} = \frac{1025}{20} = 51.25$$

[(x_i, y_i) are the various data points, for example (80, 79) for Manchester United.
n is the number of such points, in this case 20, one for each club in the Premier League.]

You will see that the point $(\bar{x}, \bar{y})$ has also been plotted on the scatter diagram and lines drawn through this point parallel to the axes. These lines divide the scatter diagram into four regions.

You can think of the point $(\bar{x}, \bar{y})$ as the middle of the scatter diagram and so treat it as a new origin. Relative to $(\bar{x}, \bar{y})$, the co-ordinates of the various points are all of the form $(x_i - \bar{x}, y_i - \bar{y})$.

In regions 1 and 3 the product $(x_i - \bar{x})(y_i - \bar{y})$ is positive for every point.

In regions 2 and 4 the product $(x_i - \bar{x})(y_i - \bar{y})$ is negative for every point.

When there is positive correlation most or all of the data points will fall in regions 1 and 3 and so you would expect the sum of these terms to be positive and large. This sum is denoted by S_{xy}

$$S_{xy} = \sum_{i=1}^{n} (x_i - \bar{x})(y_i - \bar{y})$$

When there is negative correlation most or all of the points will be in regions 2 and 4 and so you would expect the sum of these terms (S_{xy} in the equation) to be negative and large.

When there is little or no correlation the points will be scattered round all four regions. Those in regions 1 and 3 will result in positive values of $(x_i - \bar{x})(y_i - \bar{y})$ but when you add these to the negative values from the points in regions 2 and 4 you would expect most of them to cancel each other out. Consequently the total value of all the terms should be small.

By itself the actual value of S_{xy} does not tell you very much because:

(a) no allowance has been made for the number of items of data;
(b) no allowance has been made for the spread within the data;
(c) no allowance has been made for the units of x and y.

These three points may be addressed by standardising the value of $\sum_i (x_i - \bar{x})(y_i - \bar{y})$ as follows:

- Calculate $S_{xy} = \sum_i (x_i - \bar{x})(y_i - \bar{y})$

- Calculate $S_{xx} = \sum_i (x_i - \bar{x})^2$

 and $S_{yy} = \sum_i (y_i - \bar{y})^2$

- The product moment correlation coefficient r may then be found as

$$r = \frac{S_{xy}}{\sqrt{S_{xx}S_{yy}}}$$

In practice the quantities $(x_i - \bar{x})$ and $(y_i - \bar{y})$ may be difficult to work with, and so the following equivalent results are normally used:

$$S_{xx} = \sum_i x_i^2 - \frac{\left(\sum_i x_i\right)^2}{n}$$

$$S_{yy} = \sum_i y_i^2 - \frac{\left(\sum_i y_i\right)^2}{n}$$

$$S_{xy} = \sum_i x_i y_i - \frac{\left(\sum_i x_i\right)\left(\sum_i y_i\right)}{n}$$

Then

$$r = \frac{S_{xy}}{\sqrt{S_{xx}S_{yy}}}$$

The coefficient r is correctly known as *Pearson's product moment correlation coefficient*. It is often informally referred to as 'the pmcc' for short.

The quantity, r, provides a standardised measure of correlation. Its value always lies within the range −1 to +1. (If you calculate a value outside this range, you have made a mistake.) A value of +1 means perfect positive correlation; in this case all the points on a scatter diagram would lie exactly on a straight line with positive gradient. Similarly a value of −1 means perfect negative correlation. These two cases are illustrated in figure 5.10.

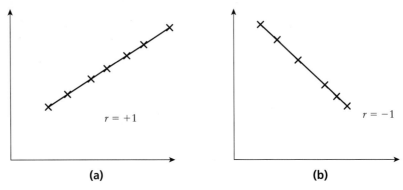

FIGURE 5.10   (a) *Perfect positive correlation;* (b) *Perfect negative correlation*

In cases of little or no correlation r takes values close to zero. The nearer the value of r is to +1 or −1, the stronger the correlation.

The calculation of Pearson's product moment correlation coefficient is shown in Example 5.1, which is worked twice using the alternative formulae.

Historical note Karl Pearson was one of the founders of modern statistics. Born in 1857, he was a man of varied interests and practised law for three years before being appointed Professor of Applied Mathematics and Mechanics at University College, London in 1884. Pearson made contributions to various branches of mathematics but is particularly remembered for his work on the application of statistics to biological problems in heredity and evolution. He died in 1936.

EXAMPLE 5.1

A gardener wishes to know if plants which produce only a few potatoes produce larger ones. He selects five plants at random, sieves out the small potatoes, counts those remaining and weighs the largest one.

Number of potatoes, x	5	5	7	8	10
Weight of largest, y (grams)	240	232	227	222	215

Calculate the correlation coefficient between x and y and comment on the result.

Product moment correlation

Solution **Method 1**

$n = 5$, $\bar{x} = 7$, $\bar{y} = 227.2$

x_i	y_i	$x_i - \bar{x}$	$y_i - \bar{y}$	$(x_i - \bar{x})^2$	$(y_i - \bar{y})^2$	$(x_i - \bar{x})(y_i - \bar{y})$
5	240	−2	12.8	4	163.84	−25.6
5	232	−2	4.8	4	23.04	−9.6
7	227	0	−0.2	0	0.04	0
8	222	1	−5.2	1	27.04	−5.2
10	215	3	−12.2	9	148.84	−36.6
$\sum$ 35	1136	0	0	18	362.80	−77.0

Then $S_{xx} = \sum_i (x_i - \bar{x})^2 = 18$

$S_{yy} = \sum_i (y_i - \bar{y})^2 = 362.80$

$S_{xy} = \sum_i (x_i - \bar{x})(y_i - \bar{y}) = -77$

∴ pmcc $r = \dfrac{S_{xy}}{\sqrt{S_{xx}S_{yy}}}$

$= \dfrac{-77}{\sqrt{18 \times 362.80}}$

$= -0.953$

Solution **Method 2**

$n = 5$

x_i	y_i	x_i^2	y_i^2	$x_i y_i$
5	240	25	57 600	1200
5	232	25	53 824	1160
7	227	49	51 529	1589
8	222	64	49 284	1776
10	215	100	46 225	2150
$\sum$ 35	1136	263	258 462	7875

Then $S_{xx} = \sum_i x_i^2 - \dfrac{\left(\sum_i x_i\right)^2}{n}$

$= 263 - \dfrac{35^2}{5}$

$= 18$

$$S_{yy} = \sum_i y_i^2 - \frac{\left(\sum_i y_i\right)^2}{n}$$

$$= 258\,462 - \frac{1136^2}{5}$$

$$= 362.8$$

$$S_{xy} = \sum_i x_i y_i - \frac{\left(\sum_i x_i\right)\left(\sum_i y_i\right)}{n}$$

$$= 7875 - \frac{35 \times 1136}{5}$$

$$= -77$$

∴ pmcc $\quad r = \dfrac{S_{xy}}{\sqrt{S_{xx} S_{yy}}}$

$$= \frac{-77}{\sqrt{18 \times 362.8}}$$

$$= -0.953$$

There is very strong negative linear correlation between the variables. Large potatoes seem to be associated with small crop sizes.

Exercise 5B

1.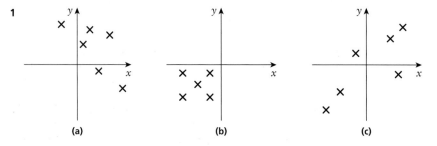

   (a)   (b)   (c)

   Three sets of bivariate data have been plotted on scatter diagrams, as illustrated. In each diagram the product moment correlation coefficient takes one of the values −1, −0.8, 0, 0.8, 1. State the appropriate value of the correlation coefficient corresponding to the scatter diagrams **(a)**, **(b)** and **(c)**.

   [Cambridge]

In questions 2–6 find Pearson's product moment correlation coefficient, r, for a number of bivariate samples.

The purpose of doing this on paper is to familiarise yourself with the routines involved. Most statisticians would actually do such calculations using a computer package, spreadsheet or a good calculator, and you should learn how to do this as well.

EXERCISE 5B

2

x	2	6	7	10
y	13	8	9	6

3

x	10	11	12	13	14	15	16	17
y	19	16	28	20	31	19	32	35

4

x	0	1	4	3	2
y	11	8	5	4	7

5

x	12	14	14	15	16	17	17	19
y	86	90	78	71	77	69	80	73

6

x	56	78	14	80	34	78	23	61
y	45	34	67	70	42	18	25	50

7 A sports reporter believes that those who are good at the high jump are also good at the long jump, and vice versa. He collects data on the best performances of nine athletes, as follows.

Athlete	A	B	C	D	E	F	G	H	I
High jump, x (metres)	2.0	2.1	1.8	2.1	1.8	1.9	1.6	1.8	1.8
Long jump, y (metres)	8.0	7.6	6.4	6.8	5.8	8.0	5.5	5.5	6.6

(a) Calculate the product moment correlation coefficient.
(b) Comment on the result.

8 It is widely believed that those who are good at chess are good at bridge, and vice versa. A commentator decides to test this theory using as data the grades of a random sample of eight people who play both games.

Player	A	B	C	D	E	F	G	H
Chess grade	160	187	129	162	149	151	189	158
Bridge grade	75	100	75	85	80	70	95	80

Calculate the product moment correlation coefficient.

9 A biologist believes that a particular type of fish develops black spots on its scales in water that is polluted by certain agricultural fertilisers. She catches a number of fish; for each one she counts the number of black spots on its scales and measures the concentration of the pollutant in the water it was swimming in. She uses these data to test for positive linear correlation between the number of spots and the level of pollution.

Fish	A	B	C	D	E	F	G	H	I	J
Pollutant concentration (parts per million)	124	59	78	79	150	12	23	45	91	68
No. of black spots	15	8	7	8	14	0	4	5	8	8

Calculate the product moment correlation coefficient.

10 The teachers at a school have a discussion as to whether girls in general run faster or slower as they get older. They decide to collect data for a random sample of girls the next time the school cross country race is held (which everybody has to take part in). They collect the following data, with the times given in minutes and the ages in years (the conversion from months to decimal parts of a year has already been carried out).

Age	Time	Age	Time	Age	Time
11.6	23.1	18.2	45.0	13.9	29.1
15.0	24.0	15.4	23.2	18.1	21.2
18.8	45.0	14.4	26.1	13.4	23.9
16.0	25.2	16.1	29.4	16.2	26.0
12.8	26.4	14.6	28.1	17.5	23.4
17.6	22.9	18.7	45.0	17.0	25.0
17.4	27.1	15.4	27.0	12.5	26.3
13.2	25.2	11.8	25.4	12.7	24.2
14.5	26.8				

(a) Calculate the product moment correlation coefficient.
(b) Plot the data on a scatter diagram and identify any outliers. Explain how they could have arisen.

Interpreting correlation

You need to be on your guard against drawing spurious conclusions from high correlation coefficients.

Correlation does not imply causation

Figures for the years 1985–93 show a high correlation between the sales of personal computers and those of microwave ovens. There is of course no direct connection between the two variables. You would be quite wrong to conclude that buying a microwave oven predisposes you to buy a computer as well, or vice versa.

Although there may be a high level of correlation between variables *A* and *B* it does not mean that *A* causes *B* or that *B* causes *A*. It may well be that a third variable *C* causes both *A* and *B*, or it may be that there is a more complicated set of relationships. In the case of personal computers and microwaves, both are clearly caused by the advance of modern technology.

Non-linear correlation

A low value of *r* tells you that there is little or no *linear* correlation. There are other forms of correlation as illustrated in figure 5.11.

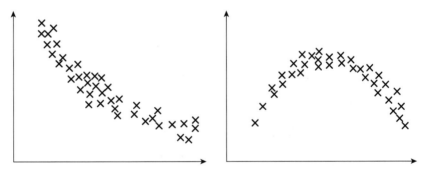

FIGURE 5.11 *Scatter diagrams showing non-linear correlation*

These diagrams show that there is an association between the variables, but not one of linear correlation.

Extrapolation

A linear relationship established over a particular domain should not be assumed to hold outside this range. For instance, there is strong correlation between the age in years and the 100-metre times of female athletes between the ages of 10 and 20 years. To extend the connection, as shown in figure 5.12 (overleaf), would suggest that veteran athletes are quicker than athletes who are in their prime and, if they live long enough, can even run 100 metres in no time at all!

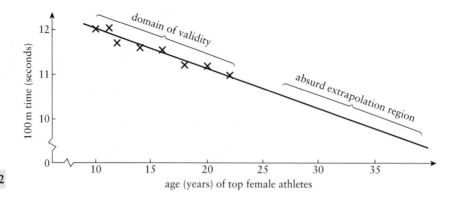

FIGURE 5.12

THE LEAST SQUARES REGRESSION LINE

A correlation coefficient provides you with a measure of the level of association between the two variables in a bivariate distribution.

If this indicates that there is a relationship, your question will be 'What is it?' In the case of linear correlation it can be expressed algebraically as a linear equation or geometrically as a straight line on the scatter diagram.

At the start of this chapter you saw how to draw a line of best fit through a set of points on a scatter diagram by eye. You did this by drawing it through the mean point, leaving roughly the same number of points above and below the line. This is obviously a very imprecise method.

Before you do any calculations you first need to look carefully at the two variables that give rise to your data. It is normal practice to plot the *dependent variable* on the vertical axis and the *independent variable* on the horizontal axis. In the example which follows, the independent variable is the time at which measurements are made. (Notice that this is a non-random variable.) The procedure leads to the equation of the *regression line*, the line of best fit in these circumstances.

Look at the scatter diagram (figure 5.13) showing the n points $A(x_1, y_1)$, $B(x_2, y_2), \ldots, N(x_n, y_n)$. On it is marked a possible line of best fit l. If the line l passed through all points there would be no problem since there would be perfect linear correlation. It does not, of course, pass though all the points and you would be very surprised if such a line did in any real situation.

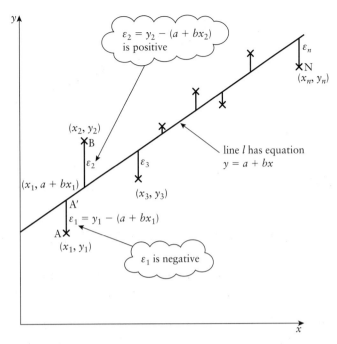

FIGURE 5.13 *Bivariate data plotted on a scatter diagram with the regression line, l, y = a + bx, and the residuals $\varepsilon_1, \varepsilon_2, \ldots, \varepsilon_n$*

By how much is it missing the points? The answer to that question is shown by the vertical lines from the points to the line. Their lengths $\varepsilon_1, \varepsilon_2, \ldots, \varepsilon_n$ are called the *residuals* and represent the variation which is not explained by the line *l*. The *least squares regression line* is the line which produces the least possible value of the sum of the squares of the residuals, $\varepsilon_1^2 + \varepsilon_2^2 + \cdots + \varepsilon_n^2$.

If the equation of the line *l* is $y = a + bx$, then it is easy to see that the point A′ on the diagram, directly above A, has co-ordinates $(x_1, a + bx_1)$ and so the corresponding residual, ε_1, is given by $\varepsilon_1 = y_1 - (a + bx_1)$. Similarly for $\varepsilon_2, \varepsilon_3, \ldots, \varepsilon_n$.

The problem is to find the values of the constants *a* and *b* in the equation of the line *l* which make $\varepsilon_1^2 + \varepsilon_2^2 + \cdots + \varepsilon_n^2$ a minimum for any particular set of data.

It may be shown that this occurs when the gradient *b* takes the value

$$b = \frac{S_{xy}}{S_{xx}}$$

and the intercept *a* is such that the line passes through the mean point $(\bar{x}, \bar{y})$. The intercept may thus be found from $\bar{y} = a + b\bar{x}$, in other words

$$a = \bar{y} - b\bar{x}$$

5 Correlation and regression

Notes

1. In the preceding work you will see that only variation in the *y* values has been considered. The reason for this is that the *x* values represent a non-random variable. That is why the residuals are vertical and not in any other direction. Thus $y_1, y_2, \ldots$ are values of a random variable Y given by $Y = a + bx + \varepsilon$ where ε is the residual variation, the variation that is not explained by the regression line.

2. The goodness of fit of a regression line may be judged by eye by looking at a scatter diagram. An informal measure which is often used is the coefficient of determination, r^2, which measures the proportion of the total variation in the dependent variable, Y, which is accounted for by the regression line. There is no standard hypothesis test based on the coefficient of determination.

3. This form of the regression line is often called the *y on x regression line*. If for some reason you had *y* as your independent variable, you would use the '*x* on *y*' form obtained by interchanging *x* and *y* in the equation.

4. Although the result above is true for a random variable on a non-random variable, it happens that, for quite different reasons, the same form of the regression line applies if both variables are random and normally distributed. If this is the case the scatter diagram will usually show an approximately elliptical distribution. Since this is a common situation, this form of the regression line may be used more widely than might at first have seemed to be the case.

EXAMPLE 5.2

A patient is given a drip feed containing a particular chemical and its concentration in his blood is measured, in suitable units, at one hour intervals for the next six hours. The doctors believe the figures to be subject to random errors, arising from both the sampling procedure and the subsequent chemical analysis, but that a linear model is appropriate.

Time, x (hours)	0	1	2	3	4	5	6
Concentration, y	2.4	4.3	5.0	6.9	9.1	11.4	13.5

(a) Find the equation of the regression line of *y* upon *x*.
(b) Estimate the concentration of the chemical in the patient's blood 3 hours and 30 minutes after treatment started.

The least squares regression line

Solution **(a)**

x_i	y_i	x_i^2	y_i^2	$x_i y_i$
0	2.4	0	5.76	0
1	4.3	1	18.49	4.3
2	5.0	4	25.00	10.0
3	6.9	9	47.61	20.7
4	9.1	16	82.81	36.4
5	11.4	25	129.96	57.0
6	13.5	36	182.25	81.0
$\sum$ 21	52.6	91	491.88	209.4

$n = 7$

$$\bar{x} = \frac{\sum x_i}{n} = \frac{21}{7} = 3 \qquad \bar{y} = \frac{\sum y_i}{n} = \frac{52.6}{7} = 7.514$$

$$S_{xx} = \sum_i x_i^2 - \frac{\left(\sum_i x_i\right)^2}{n}$$

$$= 91 - \frac{21^2}{7}$$

$$= 28$$

$$S_{xy} = \sum_i x_i y_i - \frac{\left(\sum_i x_i\right)\left(\sum_i y_i\right)}{n}$$

$$= 209.4 - \frac{21 \times 52.6}{7}$$

$$= 51.6$$

$$\therefore \quad b = \frac{S_{xy}}{S_{xx}}$$

$$= \frac{51.6}{28}$$

$$= 1.843 \text{ (3 dp)}$$

Then $a = \bar{y} - b\bar{x}$
$= 7.514 - 1.843 \times 3$
$= 1.986$

Thus the required regression line is

$$y = 1.843x + 1.986$$

(b) When $x = 3.5$, $y = 1.843 \times 3.5 + 1.986$
$= 8.4 \text{ (1 dp)}$

At time 3 hours 30 minutes the concentration is estimated to be 8.4 units.

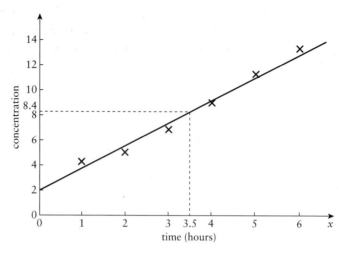

FIGURE 5.14

Note The concentration of 8.4 lies between the measured values of 6.9 at time 3 hours and 9.1 at time 4 hours and so seems quite reasonable.

EXERCISE 5C

1 For the following bivariate data obtain the equation of the least squares regression line of y on x. Estimate the value of y when $x = 12$.

x	5	10	15	20	25
y	30	28	27	27	21

2 Calculate the equation of the regression line of y on x for the following distribution and use it to estimate the value of y when $x = 42$.

x	25	30	35	40	45	50
y	78	70	65	58	48	42

3 The 1980 and 2000 catalogue prices in pence of five British postage stamps are as follows:

1980 price, x	10	20	30	40	50
2000 price, y	100	215	280	360	450

(a) Plot these data on a scatter diagram.
(b) Calculate the equation of the regression line and draw it accurately on your scatter diagram.
(c) Another stamp was valued at £5 in 1980 and at £62 in 2000. Comment.

EXERCISE 5C

4 In an investigation of the genus *Tamarix* (a shrub able to withstand drought), research workers in Tunisia measured the average vigour y (defined as the average width in centimetres of the last two annual rings) and stem density x (defined as the number of stems per m^2) at ten sites with the following results:

x	4	5	6	9	14	15	15	19	21	22
y	0.75	1.20	0.55	0.60	0.65	0.55	0	0.35	0.45	0.40

(a) Draw a scatter diagram for these data.
(b) Given that $\sum x = 130$, $\sum x^2 = 2090$, $\sum y = 5.5$ and $\sum xy = 59.95$, find the regression line of y on x, and plot it on your diagram.
(c) Use your line to estimate the average vigour for a stem density of 17 stems per m^2.
(d) Give one reason why it would be invalid to use your regression line to estimate average vigour for stem densities substantially greater than 22 stems per m^2 and explain how your regression line confirms that your reason is correct.

[MEI]

5 Each of a group of 12 apple trees is given the same spraying treatment against codling moth larvae. After the crop is gathered, the apples are examined for grubs with the following results.

Tree number	1	2	3	4	5	6	7	8	9	10	11	12
Size of crop x (hundreds of apples)	8	6	11	22	14	17	18	24	19	23	26	40
Percentage y of apples with grubs	59	58	56	53	50	45	43	42	39	38	30	27

Plot the data on a scatter diagram.

Given that $\sum x = 228$, $\sum x^2 = 5256$, $\sum y = 540$ and $\sum xy = 9324$, calculate the line of regression of y on x, and plot the line on your diagram. Estimate to the nearest whole number the expected percentage of apples with grubs for a tree carrying 2000 apples. (You may assume that these procedures are justified because x and y are jointly distributed in a suitable way.)

A second group of 12 apple trees is given a different treatment and the results are given below.

x	15	15	12	26	18	12	8	38	26	19	29	22
y	52	46	38	37	37	37	34	25	22	22	20	14

Plot the data from the second group on the same scatter diagram using a different symbol from that used before. *Without further calculation,* summarise the differences between the results of the two treatments.

[MEI]

6 The speed of a car, v metres per second, at time t seconds after it starts to accelerate is shown in the table below, for $0 \leq t \leq 10$.

t	0	1	2	3	4	5	6	7	8	9	10
v	0	3.0	6.8	10.2	12.9	16.4	20.0	21.4	23.0	24.6	26.1

$\left[\sum t = 55, \sum v = 164.4, \sum t^2 = 385, \sum v^2 = 3267.98, \sum tv = 1117.0.\right]$

The relationship between t and v is initially modelled by using all the data above and calculating a single regression line.

(a) Plot a scatter diagram of the data, with t on the horizontal axis and v on the vertical axis.

(b) Using all the data given, calculate the equation of the regression line of v on t. Give numerical coefficients in your answers correct to 3 significant figures.

(c) Calculate the product moment correlation coefficient for the given data.

(d) Comment on the validity of modelling the data by a single straight line and on the answer obtained in part (c).

[Cambridge]

7 The table below shows the names of five toy construction kits which were bought from a catalogue, the numbers of pieces, n, found in each and the corresponding prices paid, £p.

Name	Set 1	Set 3	Set 4	Set 5	Set 6
n	11	21	28	37	75
p	11	26	34	41	88

$\left[\sum n = 172, \sum p = 200, \sum n^2 = 8340, \sum p^2 = 11\,378, \sum np = 9736.\right]$

(a) Plot a scatter diagram of the data, with n on the horizontal axis and p on the vertical axis.

(b) Calculate the equation of the regression line of p on n, and plot this line on your scatter diagram. Use your equation to estimate the price of Set 2, which is not listed in the catalogue, but is thought to have 15 pieces. Give your answer correct to the nearest pound.

(c) Calculate the product moment correlation coefficient for the given data, giving your answer correct to 3 decimal places, and interpret the result in terms of your scatter diagram.

[Cambridge]

Exercise 5C

8 The results of an experiment to determine how the percentage sand content of soil y varies with depth in centimetres below ground level x are given in the following table.

x	0	6	12	18	24	30	36	42	48
y	80.6	63.0	64.3	62.5	57.5	59.2	40.8	46.9	37.6

Calculate
(a) the product moment correlation coefficient of x and y
(b) the equation of the line of regression of y on x.

[MEI, part]

9 Madame Zora predicted when 13 women would give birth to their first child. The predicted and actual ages are shown below.

Person	A	B	C	D	E	F	G	H	I	J	K	L	M
Predicted age x (years)	24	30	28	36	20	22	31	28	21	29	40	25	27
Actual age y (years)	23	31	28	35	20	25	45	30	22	27	40	27	26

(a) Draw a scatter diagram of these data.
(b) Calculate the equation of the regression line of y on x and draw this line on the scatter diagram.
(c) Comment upon the results obtained, particularly in view of the data for person G. What further action would you suggest?

10 Observations of a cactus graft were made under controlled environmental conditions. The table gives the observed heights y cm of the graft at x weeks after grafting. Also given are the values of $z = \ln y$.

x	1	2	3	4	5	6	8	10
y	2.0	2.4	2.5	5.1	6.7	9.4	18.3	35.1
$z = \ln y$	0.69	0.88	0.92	1.63	1.90	2.24	2.91	3.56

(a) Draw two scatter diagrams, one for y and x, and one for z and x.
(b) It is desired to estimate the height of the graft seven weeks after grafting. Explain why your scatter diagrams suggest the use of the line of regression of z on x for this purpose, but not the line of regression of y on x.
(c) Obtain the required estimate given that $\sum x = 39$, $\sum x^2 = 255$, $\sum z = 14.73$, $\sum z^2 = 34.5231$, $\sum xz = 93.55$.

[MEI]

LINEAR CODING

Some questions on correlation and regression present data in a form which has been coded, i.e. scaled in a linear way.

This does not affect the value of the pmcc at all, for it turns out that the pmcc of the coded data is identical in value to the pmcc for the raw, uncoded data. This important principle may be summarised as follows:

> The value of the product moment correlation coefficient r is independent of the scale of measurement.

For a regression line, the best approach is to establish the equation connecting the coded variables first. Then substitute in the expressions representing the coding.

These two techniques are illustrated in Example 5.3. Further examples may be found in Exercise 5D which follows at the end of this chapter.

EXAMPLE 5.3

The marks scored by a random sample of 15 A level candidates in a Pure Mathematics 1 examination (p) and a Statistics 1 examination (s) are summarised as follows:

$n = 15$, $\sum (p - 40) = -71$, $\sum (s - 40) = 43$, $\sum (p - 40)^2 = 6975$,
$\sum (s - 40)^2 = 5511$, $\sum (p - 40)(s - 40) = 4955$.

(a) The value of $\sum (p - 40)$ is negative while that of $\sum (s - 40)$ is positive. What does this suggest about the relative abilities of the candidates in Pure Mathematics 1 and Statistics 1?
(b) Calculate the pmcc for this data.
(c) Obtain the equation of the regression line of s on p.
(d) Use your equation from (c) to predict the Statistics 1 score for a candidate who scores 70 (the maximum under this particular system) in Pure Mathematics 1. Does this appear to be a realistic prediction?

Solution

(a) $\sum (p - 40)$ negative $\Rightarrow$ mean P1 score is less than 40.
$\sum (s - 40)$ positive $\Rightarrow$ mean S1 score is more than 40.

This suggests that the candidates are tending to get a higher score in Statistics 1 than in Pure Mathematics 1.

(b) Let $x = p - 40$, $y = s - 40$.
Then $n = 15$, $\sum x = -71$, $\sum y = 43$, $\sum x^2 = 6975$, $\sum y^2 = 5511$, $\sum xy = 4955$.

$$S_{xx} = \sum x^2 - \frac{(\sum x)^2}{n}$$

$$= 6975 - \frac{(-71)^2}{15}$$

$$= 6638.9\dot{3}$$

$$S_{yy} = \sum y^2 - \frac{(\sum y)^2}{n}$$

$$= 5511 - \frac{43^2}{15}$$

$$= 5387.7\dot{3}$$

$$S_{xy} = \sum xy - \frac{(\sum x)(\sum y)}{n}$$

$$= 4955 - \frac{(-71) \times 43}{15}$$

$$= 5158.5\dot{3}$$

Then $r = \dfrac{S_{xy}}{\sqrt{S_{xx} S_{yy}}}$

$$= \frac{5158.53}{\sqrt{6638.93 \times 5387.73}}$$

$$= 0.8625$$

(c) Let $y = a + bx$

Then $b = \dfrac{S_{xy}}{S_{xx}}$

$$= \frac{5158.53}{6638.93}$$

$$= 0.7770$$

Thus $a = \bar{y} - b\bar{x}$

$$= \frac{43}{15} - 0.7770 \times \frac{(-71)}{15}$$

$$= 6.5445$$

The regression line of y on x is then

$$y = 6.5445 + 0.7770x$$

This becomes

$$s - 40 = 6.5445 + 0.7770(p - 40)$$

which rearranges to give

$$s = 15.4645 + 0.7770p$$

(d) When $p = 70$

$$s = 15.46 + 0.777p$$

$$= 69.85$$

$$= 70 \text{ (nearest whole number)}$$

This does look realistic, since we are predicting that a candidate who achieves full marks in Pure Mathematics 1 (which may be a little harder) also achieves full marks in Statistics 1 (which may be a little easier).

5 Correlation and regression

Exercise 5D Examination style questions

1. A random sample of students who are shortly to sit an examination are asked to keep a record of how long they spend revising, in order to investigate whether more revision time is associated with a higher mark. The data are given below, with x hours being the revision time (correct to the nearest $\frac{1}{2}$ hour) and $y\%$ being the mark scored in the examination.

x	0	3	4.5	3.5	7	5.5	5	6.5	6	10.5	2
y	36	52	52	57	60	61	63	63	64	70	89

$n = 11 \quad \sum x = 53.5 \quad \sum y = 667 \quad \sum x^2 = 338.25 \quad \sum y^2 = 42\,129 \quad \sum xy = 3366.5$

(a) Obtain the value of the product moment correlation coefficient for the data.
(b) Without further calculation, state the effect of the data $x = 2$, $y = 89$ on the value of the product moment correlation coefficient. Explain whether or not this point should be excluded when carrying out the hypothesis test.

[MEI, part]

2. A car manufacturer is testing the braking distance for a new model of car. The table shows the braking distance, y metres, for different speeds, x km hr^{-1}, when the brakes were applied.

Speed of car, x km hr^{-1}	30	50	70	90	110	130
Braking distance, y metres (to the nearest 5 metres)	25	50	85	155	235	350

$\sum x = 480 \quad \sum x^2 = 45\,400 \quad \sum y = 900 \quad \sum y^2 = 212\,100 \quad \sum xy = 94\,500$

(a) Plot a scatter diagram to illustrate the data.
(b) Calculate the equation of the regression line of y on x and draw the line on your scatter diagram.
(c) Use your regression equation to predict values of y when $x = 100$ and $x = 150$. Comment, with reasons, on the likely accuracy of these predictions.
(d) Discuss briefly whether the regression line provides a good model or whether there is a better way of modelling the relationship between y and x.

[MEI]

3. An experiment was conducted to determine the mass, y g, of a chemical that would dissolve in 100 ml of water at $x\,°C$. The results of the experiment were as follows.

Temperature ($x\,°C$)	10	20	30	40	50
Mass (y g)	61	64	70	73	75

(a) Represent the data on graph paper.

(b) Calculate the equation of the regression line of y on x. Draw this line on your graph.

(c) Calculate an estimate of the mass of the chemical that would dissolve in the water at 35 °C.

Suggest a range of temperatures for which such estimates are reliable. Give a reason for your answer.

(d) Calculate the residuals for each of the temperatures. Illustrate them on your graph.

(e) The regression line is often referred to as 'the least squares regression line'. Explain what this means in relation to the residuals.

[MEI]

4 The bivariate sample illustrated in the scatter diagram shows the heights, x cm, and masses, y kg, of a random sample of 20 students.

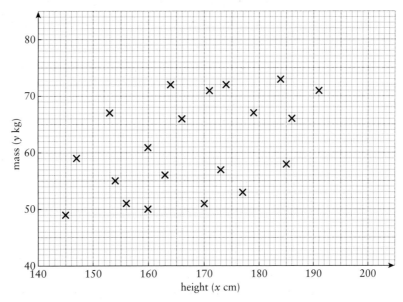

You are given that $\sum x = 3358$, $\sum x^2 = 567\,190$, $\sum y = 1225$, $\sum y^2 = 76\,357$, $\sum xy = 206\,680$.

(a) Calculate the product moment correlation coefficient.

(b) What feature of the scatter diagram suggests that the pmcc is an appropriate statistic?

(c) A statistics student suggests that a positive correlation between height and mass implies that 'the taller a student is the heavier he or she will be'. Comment on this statement with reference to your conclusions in part (b).

[MEI, adapted]

5. A farmer is investigating the relationship between the density at which a crop is planted and the quality. By using more intensive methods he can increase the yield, but he suspects that the percentage of high quality produce may fall. The farmer collects the following data.

Seed per acre (x) in suitable units	120	130	140	150	160	170
Percentage of high quality produce (y)	31.3	28.9	25.4	21.3	21.0	10.7

$n = 6$ $\sum x = 870$ $\sum x^2 = 127\,900$
$\sum xy = 19\,443$ $\sum y = 138.6$ $\sum y^2 = 3469.24$

(a) Draw a sketch to illustrate the data. Hence discuss how suitable a straight line model would be for the relationship between y and x.

(b) Calculate the equation of the regression line of y on x.

(c) Obtain from your regression line the predicted values of y at $x = 145$ and $x = 180$. Comment, with reasons, on the likely accuracy of these predictions.

[MEI]

6. An experiment was conducted to see whether there was any relationship between the maximum tidal current, y cm s^{-1}, and the tidal range, x metres, at a particular marine location. (The *tidal range* is the difference between the height of high tide and the height of low tide.) Readings were taken over a period of 12 days, and the results are shown in the following table.

x	2.0	2.4	3.0	3.1	3.4	3.7	3.8	3.9	4.0	4.5	4.6	4.9
y	15.2	22.0	25.2	33.0	33.1	34.2	51.0	42.3	45.0	50.7	61.0	59.2

$\left[\sum x = 43.3, \sum y = 471.9, \sum x^2 = 164.69, \sum y^2 = 20\,915.75, \sum xy = 1837.78.\right]$

The scatter diagram below illustrates the data.

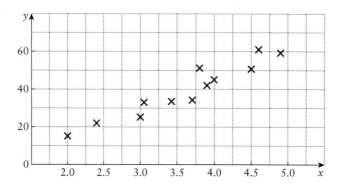

(a) Calculate the product moment correlation coefficient for the data, and comment briefly on your answer with reference to the appearance of the scatter diagram.

EXERCISE 5D

(b) Calculate the equation of the regression line of maximum tidal current on tidal range.

(c) Estimate the maximum tidal current on a day when the tidal range is 4.2 m, and indicate briefly how reliable an estimate you consider your answer to be.

(d) It is suggested that the equation found in part (b) could be used to predict the maximum tidal current on a day when the tidal range is 15 m. Comment briefly on the validity of this suggestion.

[OCR]

7 In 1980 a river was restocked with fish. The local angling club kept a record of the number of fish, y, caught on a stretch of the river t years after restocking. Some of the records are given in the table below.

t	1	2	3	4	5	7	10	12
y	185	170	172	166	164	159	162	157

(a) Plot a scatter diagram of y against t.

A local conservation organisation believed that a linear regression of the form $y = p + qt$ could be used to model this situation.

(b) Give an interpretation of q.

The conservation organisation found the equation of the regression line to be

$$y = 177.28 - 1.89t.$$

(c) Explain the long term implications of this model for the number of fish caught.

A member of the angling club suggests an alternative model based on introducing the new variable $x = \frac{1}{t}$. The following statistics are calculated:

$$\sum x = 2.6095, \sum y = 1335, \sum x^2 = 1.5010, \sum xy = 453.6310.$$

(d) Calculate the regression equation of y on x in the form $y = a + bx$.

(e) Explain the long term implications of his new model for the number of fish caught.

(f) Give an interpretation of the value of a.

[Edexcel]

8 In an experiment to discover whether the amount of 'brown fat' in an animal's body affects its tendency to obesity, a biologist took a number of rats from a warm room and put them into a very cold room. The time, in seconds, taken for each to start shivering (which is related to the amount of brown fat in the body) was recorded. Nine of these rats were selected, and allowed unlimited access to food for four weeks, after which the increase in their body weight, in kg, was measured. The results are recorded below.

Time to shiver, t	10	10	14	15	21	35	35	47	50
Weight gain, w	0.00	−0.02	0.04	0.06	0.10	0.15	0.18	0.18	0.19

You may assume that
$$\sum t = 237, \sum t^2 = 8221, \sum tw = 32.87, \sum w = 0.88.$$

(a) Plot a graph of w against t.
(b) Find S_{tt} and S_{tw}.
(c) Calculate the equation of the regression line of w on t in the form $w = a + bt$.
(d) Draw the regression line on your graph.
(e) Give an interpretation of the values of a and b in this context.

On looking at the graph, the biologist conducting the experiment felt that the model was a reasonable one, but that it could be improved.

(f) Give a reason why the model seems a reasonable one (no calculation is required), and suggest a reason why the biologist thought it could be improved.

A tenth rat started to shiver 180 seconds after it was placed in the cold room.

(g) Determine the prediction for weight gain given by the model, and state, with a reason, whether this prediction is reliable.

Some rats do not shiver at all while in the cold room.

(h) State how this poses a problem when using the linear model.

[Edexcel]

9 A newspaper printed an article with the headline 'Rural Population Goes Downhill'. The article contained the statement 'The population in this region has been shown statistically to be falling steadily in the period from 1960 to 1984.' The actual data on which the article was based are shown in the table.

Year, t	1960	1962	1964	1966	1968	1970	1972
Population, p	460 905	460 972	460 776	460 895	460 706	460 648	460 577

Year, t	1974	1976	1978	1980	1982	1984
Population, p	460 668	460 592	460 660	460 512	460 586	460 352

Exercise 5D

You may assume that

$$\sum (t - 1970) = 26, \sum (t - 1970)^2 = 780,$$
$$\sum (t - 1970)(p - 460\,500) = -9736, \sum (p - 460\,500) = 2349,$$
$$\sum (p - 460\,500)^2 = 781\,011.$$

A student desires to examine the newspaper's statement.

(a) Calculate the product moment correlation coefficient for the data.

(b) State one feature of the data which does not support the newspaper's statement.

[Edexcel, part]

10 In a physics experiment, a bottle of milk was brought from a cool room into a warm room. Its temperature, $y\,°C$, was recorded at t minutes after it was brought in, for 11 different values of t. The results are summarised as:

$$\sum t = 44, \sum t^2 = 180.4, \sum ty = 824.5, \sum y = 205.$$

(a) Calculate the equation of the line of regression of y on t in the form $y = a + bt$.

(b) Explain the practical significance of the value of a.

(c) Use your equation to estimate the values of y at $t = 4.5$ and $t = 20.0$.

(d) State, with a reason, which of these estimates is likely to be the more reliable.

The experimenter plotted a graph of y against t, but used only the data in the table below.

Time (minutes), t	3.0	3.4	3.8	4.2	4.6	5.0
Temperature (°C), y	17.0	18.3	18.6	18.9	19.3	19.4

(e) Plot this graph, and on it draw the line of regression.

(f) State why the linear model could not be valid for very large values of the time.

(g) Using your graph, comment on whether the model is a reasonable one, and state, giving a reason, whether you consider that a more refined model could be found.

[Edexcel]

KEY POINTS

1. Initial evidence for linear correlation should be determined by plotting a scatter graph.

2. Linear correlation is measured by the product moment correlation coefficient, or pmcc, r.

3. Let
$$S_{xx} = \sum x^2 - \frac{(\sum x)^2}{n}$$

$$S_{yy} = \sum y^2 - \frac{(\sum y)^2}{n}$$

$$S_{xy} = \sum xy - \frac{(\sum x)(\sum y)}{n}$$

Then
$$r = \frac{S_{xy}}{\sqrt{S_{xx}S_{yy}}}$$

4. The least squares regression line has equation $y = a + bx$ where
$$b = \frac{S_{xy}}{S_{xx}}$$

The intercept a is then found using
$$a = \bar{y} - b\bar{x}$$

where $(\bar{x}, \bar{y})$ is the mean point.

5. The value of the pmcc is independent of the scale of measurement.

Chapter six

DISCRETE RANDOM VARIABLES

An approximate answer to the right problem is worth a good deal more than an exact answer to an approximate problem.

John Tukey

The number of babies at any pregnancy is an example of a *discrete random variable*. It is *discrete* because it takes only particular values, 1, 2, 3, 4, etc.; you cannot have 2.4 or 0.315 babies. A discrete variable does not, in general, have to be a positive whole number. The shoe size of a set of students is a discrete variable taking the possible values ... 5, $5\frac{1}{2}$, 6, $6\frac{1}{2}$, ... By contrast a baby's birth weight is a *continuous* variable which can take any value between two sensible limits. Typical continuous variables are people's heights, weights of elephants and the marathon times of a set of runners.

The number of babies born at any pregnancy is *random* because on becoming pregnant a woman cannot determine how many babies she is going to have. Having twins or triplets is a chance event. (The effects of fertility drugs and in vitro fertilisation have been ignored.)

It is a *variable* because it takes different numerical values.

NOTATION

A discrete random variable is usually denoted by an upper case letter, such as X, Y, or Z, etc. You may think of this as the name of the variable. The particular values the variable takes are denoted by lower case letters, such as x, y, z. Sometimes these are given suffices x_1, x_2, x_3 etc. Thus $P(X = x_1)$ means 'the probability that discrete random variable X takes the value x_1.'

THE CONDITIONS FOR A DISCRETE RANDOM VARIABLE

If the outcome of a process can be stated as a number, X, which can take different possible values $x_1, x_2, \ldots, x_n$ at random, then X is a discrete random variable. The probabilities $p_1, p_2, \ldots, p_n$ of the different values must sum to 1.

Outcome	x_1	x_2	x_3	$\ldots$	x_n
Probability	p_1	p_2	p_3	$\ldots$	p_n

$$p_1 + p_2 + p_3 + \ldots + p_n = 1 \qquad p_i \geq 0, \text{ for } i = 1, \ldots, n$$

Another way of saying this is that the various outcomes cover all possibilities, that is they are *exhaustive*.

Note   Since p_1, p_2 etc. are probabilities, none of them may exceed 1.

This gives the probability distribution of X. The rule which assigns probabilities to the various outcomes is known as the *probability function* of X. Sometimes it is possible to write the probability distribution as a formula.

DIAGRAMS OF PROBABILITY DISTRIBUTIONS

If you wish to draw a diagram to show the distribution of a discrete random variable, you must show clearly that the variable is indeed discrete, as in the two diagrams of figure 6.1.

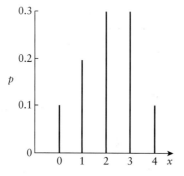
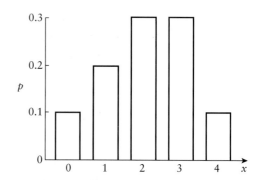

FIGURE 6.1

This may be contrasted with the distribution of a *continuous* variable, which is drawn as a continuous curve, as in figure 6.2. Its equation is called the *probability density function* and usually written f(x). The area under this curve represents probability.

Diagrams of probability distributions

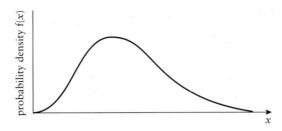

FIGURE 6.2

You will often find yourself wanting to know the probability that a random variable takes a value no more than a certain number, $P(X \leq x)$. This is known as the *cumulative distribution function* and generally denoted by $F(x)$; see figure 6.3.

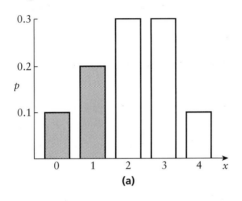
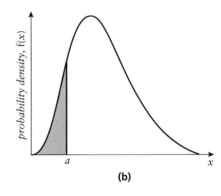

FIGURE 6.3 *The shaded area is the cumulative distribution function:* **(a)** $F(1)$; **(b)** $F(a)$

EXAMPLE 6.1

The variable X is the score on an unbiased die after it is rolled.
(a) Write down the probability distribution of X.
(b) Show that X satisfies the conditions for it to be a discrete random variable.

Solution (a) The probability distribution of X is:

Outcome	1	2	3	4	5	6
Probability	$\frac{1}{6}$	$\frac{1}{6}$	$\frac{1}{6}$	$\frac{1}{6}$	$\frac{1}{6}$	$\frac{1}{6}$

Notice that $P(X = 1) + P(X = 2) + \cdots + P(X = 6) = 1$
because $\frac{1}{6} + \frac{1}{6} + \cdots + \frac{1}{6} = 1$.

(b) The set of possible values of X, $\{1, 2, 3, 4, 5, 6\}$, forms a discrete set. The outcome from rolling a die is clearly random.

Therefore X is a discrete random variable.

EXAMPLE 6.2

A card is selected at random from a normal pack of 52 playing cards. If it is a spade the experiment ends, otherwise it is replaced, the pack is shuffled and another card is selected. What is the probability distribution of X, the number of cards selected up to and including the first spade?

Solution This situation is shown in the tree diagram below. You will see that there is no limit to the value which X may take, although larger values of X become more and more unlikely.

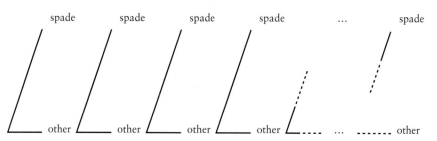

FIGURE 6.4

x	1	2	3	4	...	r
$P(X = x)$	$\frac{1}{4}$	$\frac{3}{4} \times \frac{1}{4}$	$\left(\frac{3}{4}\right)^2 \times \frac{1}{4}$	$\left(\frac{3}{4}\right)^3 \times \frac{1}{4}$	...	$\left(\frac{3}{4}\right)^{r-1} \times \frac{1}{4}$

So the probability distribution is given by

$$P(X = r) = \left(\tfrac{3}{4}\right)^{r-1} \times \left(\tfrac{1}{4}\right) \quad \text{for } r = 1, 2, 3, \ldots$$

Check: Since this is a probability distribution, the sum of all the probabilities should be 1. Is it true that

$$\tfrac{1}{4} + \tfrac{3}{4} \times \tfrac{1}{4} + \left(\tfrac{3}{4}\right)^2 \times \tfrac{1}{4} + \left(\tfrac{3}{4}\right)^3 \times \tfrac{1}{4} + \cdots + \left(\tfrac{3}{4}\right)^{r-1} \times \tfrac{1}{4} + \cdots = 1?$$

The sequence of probabilities form a geometric distribution. You will recognise that it is an infinite geometric series, with the first term, $a = \tfrac{1}{4}$, and the common ratio, $r = \tfrac{3}{4}$.

Substituting these values in the formula for the sum of an infinite geometric series gives

$$S_\infty = \frac{a}{1-r} = \frac{\tfrac{1}{4}}{1 - \tfrac{3}{4}} = \frac{\tfrac{1}{4}}{\tfrac{1}{4}} = 1 \text{ as required.}$$

EXAMPLE 6.3

The probability distribution of a random variable Y is given by

$$P(Y = y) = cy \quad \text{for } y = 1, 2, 3, 4, 5.$$

(a) Given that c is a constant find the value of c.
(b) Hence find the probability that $Y > 3$.

Solution

(a) The relationship $P(Y = y) = cy$ gives the table

Y	1	2	3	4	5
P(Y)	c	2c	3c	4c	5c

and since Y is a random variable $c + 2c + 3c + 4c + 5c = 1$
$$15c = 1$$
$$c = \tfrac{1}{15}$$

(b) $P(Y > 3) = P(Y = 4) + P(Y = 5)$
$= \tfrac{4}{15} + \tfrac{5}{15}$
$= \tfrac{9}{15}$
$= \tfrac{3}{5}$

Note

You may have found the expression $P(Y = y)$ somewhat curious. You should read it as 'the probability that the random value Y has the particular value y'.

EXERCISE 6A

1 The probability distribution of a discrete random variable, X, is given by

$$P(X = x) = kx$$

where k is a constant, for $x = 1, 2, 3, 4$.
Find the value of k.

2 The probability that a variable, Y, takes the value y is given by

$$P(Y = y) = (\tfrac{1}{6})(\tfrac{5}{6})^{y-1} \quad \text{for } y = 1, 2, 3, \ldots$$

Show that Y satisfies the conditions for it to be a discrete random variable and suggest a situation it could be modelling.

3 The probability distribution of a discrete random variable Z is given by

$$P(Z = z) = \frac{az}{8} \quad \text{for } z = 2, 4, 6, 8.$$

(a) Find the value of a.
(b) Hence find $P(Z < 6)$.

4 The random variable X is given by the number of heads obtained when five fair coins are tossed. Write out the probability distribution for X.

5 The probability distribution of a discrete random variable Y is given by
$$P(Y = y) = cy \quad \text{for } y = 1, 2, 3, 4, 5, 6, 7$$
where c is a constant.

Find the value of c and the probability that Y is less than 3.

6 The random variable X is given by the sum of the scores when two ordinary dice are thrown.

Write out the probability distribution of X and verify that X is a discrete random variable.

7 The random variable Y is the difference of the scores when two ordinary dice are thrown.
   (a) Write out the probability distribution of Y.
   (b) Find $P(Y < 3)$.

8 The random variable Z is the number of heads obtained when four fair coins are tossed.
   (a) Write out the probability distribution of Z.
   (b) Find the probability that there are more heads obtained than tails.

9 A box contains six black and four red pens. Three pens are taken at random from the box. The random variable X is the number of red pens obtained. Find the probability distribution of X.

10 Three committee members are to be selected from six men and seven women. Write out the probability distribution for the number of men appointed to the committee assuming the selection is done at random.

11 Two tetrahedral dice each with faces labelled 1, 2, 3 and 4 are thrown and the random variable X is the product of the numbers on which the dice fall.
   (a) Find the probability distribution of X.
   (b) What is the probability that any throw of the dice results in a value of X which is an odd number?

12 Four cards are drawn, without replacement, from a normal pack of 52. Write the probability distribution for the number of red cards chosen.

13 An ornithologist carries out a study of the numbers of eggs laid per pair by a species of rare bird in its annual breeding season. He concludes that it may be considered as a discrete random variable X with probability distribution given by
$$P(X = 0) = 0.2$$
$$P(X = x) = k(4x - x^2) \quad \text{for } x = 1, 2, 3 \text{ or } 4$$
$$P(X = x) = 0 \quad \text{for } x > 4.$$

(a) Find the value of *k* and write out the probability distribution as a table.

The ornithologist observes that the probability of survival (that is of an egg hatching and of the chick living to the stage of leaving the nest) is dependent on the number of eggs in the nest. He estimates these probabilities to be as follows.

x	Probability of survival
1	0.8
2	0.6
3	0.4

(b) Find, in the form of a table, the probability distribution of the number of chicks surviving per pair of adults.

14 A sociologist is investigating the changing pattern of the numbers of children which women have in a country. She denotes the present number by the random variable X which she finds to have the following distribution.

Number of children x	0	1	2	3	4	5+
Probability $P(X = x)$	0.09	0.22	a	0.19	0.08	negligible

(a) Find the value of a.

She is anxious to find an algebraic expression for the probability distribution and tries

$$P(X = x) = k(x + 1)(5 - x) \quad \text{for } 0 \leqslant x \leqslant 5$$
$$\text{(x may only take integer values)}$$
$$= 0 \text{ otherwise}$$

(b) Find the value of k for this model.
(c) Compare the algebraic model with the probabilities she found. Do you think it is a good model? Is there any reason why it must be possible to express the probability distribution in a neat algebraic form?

15 In a game, each player throws four ordinary six-sided dice. The random variable X is the largest number showing on the dice.
(a) Find the probability that $X = 1$.
(b) Find the probability that $X \leqslant 2$ and deduce that the probability that $X = 2$ is $\frac{5}{432}$.
(c) Find the probability that $X = 3$.
(d) Find the probability that $X = 6$ and explain without further calculation why 6 is the most likely value of X.

[MEI]

EXPECTATION

If a discrete random variable, X, takes possible values $x_1, x_2, x_3, \ldots$ with associated probabilities $p_1, p_2, p_3, \ldots$, the *expectation* E(X) of X is given by

$$E(X) = \sum_i x_i p_i.$$

Note

The terms *average* and *mean* are often used to convey the same idea as expectation but this may cause confusion.

Expectation is actually the mean of the underlying distribution, the *parent population*. Expectation is denoted by μ (pronounced 'mew'). The terms average and mean can be applied to either the parent population or a particular sample.

Figures defining the parent population are called *parameters* and usually denoted by Greek letters. The letter μ is used for expectation.

EXAMPLE 6.4

What is the expectation of the score when an unbiased die is rolled once?

Solution   The probability distribution is:

Outcome	1	2	3	4	5	6
Probability	$\frac{1}{6}$	$\frac{1}{6}$	$\frac{1}{6}$	$\frac{1}{6}$	$\frac{1}{6}$	$\frac{1}{6}$

The expected score, $E(X) = \sum x_i p_i$
$= 1 \times \frac{1}{6} + 2 \times \frac{1}{6} + 3 \times \frac{1}{6} + 4 \times \frac{1}{6} + 5 \times \frac{1}{6} + 6 \times \frac{1}{6}$
$= \frac{21}{6} = 3.5$

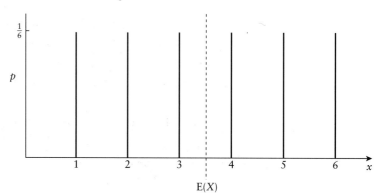

FIGURE 6.5

Notice that in this case the value of 3.5 for E(X) is an impossible outcome. This is quite often the case (e.g. an average family with 2.4 children) but it is wrong to round it to the nearest apparently sensible value. When used in statistics, expectation is a technical term, with a different meaning from that in everyday English.

EXAMPLE 6.5

A fruit machine is constructed so that the cost is 10p per turn and the amount won by the user in pence at each turn has the following probability distribution:

Payout (pence)	Nil	50	100	500
Less stake	−10	−10	−10	−10
Winnings	−10	40	90	490
Probability	0.94	0.03	0.02	0.01

(a) What is the expected loss per turn?
(b) How much can the user expect to lose after 20 turns?

Solution Let the random variable X be the amount won per turn in pence by the user.
(a) $E(X) = -10 \times 0.94 + 40 \times 0.03 + 90 \times 0.02 + 490 \times 0.01$
$= -9.4 + 7.9$
$= -1.5$
The expectation of the amount won per turn is −1.5 pence, that is a loss of 1.5 pence.
(b) After 20 turns the user can expect to lose $20 \times 1.5 = 30$ pence.

Expectation of a Function of X, $E(g[X])$

Sometimes you will need to find the expectation of a function of a random variable. That sounds rather forbidding and you may think the same of the definition given below at first sight. However, as you will see in the next two examples, the procedure is straightforward and common sense.

> If $g[X]$ is a function of the discrete random variable X then $E(g[X])$ is given by
> $$E(g[X]) = \sum_i g[x_i] P(X = x_i)$$

EXAMPLE 6.6

What is the expectation of the square of the number that comes up when a fair die is rolled?

Solution Let the random variable X be the number that comes up when the die is rolled.

$$g[X] = X^2$$

$$E(g[X]) = E(X^2) = \sum_i x_i^2 P(X = x_i)$$

$$= 1^2 \times \tfrac{1}{6} + 2^2 \times \tfrac{1}{6} + 3^2 \times \tfrac{1}{6} + 4^2 \times \tfrac{1}{6} + 5^2 \times \tfrac{1}{6} + 6^2 \times \tfrac{1}{6}$$

$$= 1 \times \tfrac{1}{6} + 4 \times \tfrac{1}{6} + 9 \times \tfrac{1}{6} + 16 \times \tfrac{1}{6} + 25 \times \tfrac{1}{6} + 36 \times \tfrac{1}{6}$$

$$= \frac{91}{6}$$

$$= 15.17$$

Note This calculation could also have been set out in table form as shown below.

x_i	$P(X = x_i)$	x_i^2	$x_i^2 P(X = x_i)$
1	$\tfrac{1}{6}$	1	$\tfrac{1}{6}$
2	$\tfrac{1}{6}$	4	$\tfrac{4}{6}$
3	$\tfrac{1}{6}$	9	$\tfrac{9}{6}$
4	$\tfrac{1}{6}$	16	$\tfrac{16}{6}$
5	$\tfrac{1}{6}$	25	$\tfrac{25}{6}$
6	$\tfrac{1}{6}$	36	$\tfrac{36}{6}$
		Σ	$\tfrac{91}{6}$

$$E(g[X]) = \tfrac{91}{6} = 15.17$$

Expectation algebra

EXAMPLE 6.7

A random variable X has the following probability distribution:

Outcome	1	2	3
Probability	0.4	0.4	0.2

(a) Calculate E(4X + 5).
(b) Calculate 4E(X) + 5.
(c) Comment on the relationship between your answers to parts (a) and (b).

Solution (a) $E(g[X]) = \sum_i g[x_i] \cdot P(X = x_i)$ with $g[X] = 4X + 5$

x_i	1	2	3
$g[x_i]$	9	13	17
$P(X = x_i)$	0.4	0.4	0.2

$$E(4X + 5) = E(g[X])$$
$$= 9 \times 0.4 + 13 \times 0.4 + 17 \times 0.2$$
$$= 12.2$$

(b) $E(X) = 1 \times 0.4 + 2 \times 0.4 + 3 \times 0.2 = 1.8$

and so

$$4E(X) + 5 = 4 \times 1.8 + 5$$
$$= 12.2$$

(c) Clearly E(4X + 5) = 4E(X) + 5, both having the value 12.2.

EXPECTATION ALGEBRA

In Example 6.7 above you found that E(4X + 5) = 4E(X) + 5.

The working was numerical, showing that both expressions came out to be 12.2, but it could also have been shown algebraically. This would have been set out as follows:

Proof	*Reasons (general rules)*
E(4X + 5) = E(4X) + E(5)	$E(X \pm Y) = E(X) \pm E(Y)$
= 4E(X) + E(5)	$E(aX) = aE(X)$
= 4E(X) + 5	$E(c) = c$

Look at the general rules on the right-hand side of the page. (X and Y are random variables, a and c are constants.) They are important but they are also common sense.

Notice the last one, which in this case means the expectation of 5 is 5. Of course it is; 5 cannot be anything else but 5. It is so obvious that sometimes people find it confusing!

These rules can be extended to take in the expectation of the sum of two functions of a random variable.

$$E(f[X] + g[X]) = E(f[X]) + E(g[X])$$

where f and g are both functions of X.

Proof
By definition

$$E(f[X] + g[X]) = \sum_i (f[x_i] + g[x_i]) \cdot P(X = x_i)$$
$$= \sum_i f[x_i] \cdot P(X = x_i) + \sum_i g[x_i] \cdot P(X = x_i)$$
$$= E(f[X]) + E(g[X])$$

EXERCISE 6B

1 Find the expectation of the number of tails when three fair coins are tossed.

2 Find the expectation for the outcome with the following distribution:

Outcome	1	2	3	4	5
Probability	0.2	0.2	0.4	0.1	0.1

3 A discrete random variable X can assume only the values 4 and 5, and has expectation 4.2. Find the two probabilities $P(X = 4)$ and $P(X = 5)$.

4 A discrete random variable Y can take only the values 50 and 100. Given that $E(Y) = 80$, write out the probability distribution of Y.

5 The probability distribution of a discrete random variable Z is given by
$$P(Z = z) = cz \quad \text{for } z = 2, 3, 4$$
where c is a constant. Find $E(Z)$.

6 A discrete random variable X has the following probability distribution:

x	-2	-1	0	1	2
$P(X = x)$	0.1	0.15	0.15	0.35	0.25

Find
(a) $P(-1 \leqslant X \leqslant 1)$
(b) $E(X)$
(c) $P(X < 0.8)$.

EXERCISE 6B

7 An unbiased tetrahedral die has faces labelled 2, 4, 6 and 8. If the die lands on the face marked 2, the player has to pay 50p. If it lands on a face marked with a 4 or a 6 the player wins 20p, and if it lands on the face labelled 8 then no money changes hands.

Find the expected gain or loss of the player after
  (a) 1 throw
  (b) 3 throws
  (c) 100 throws.

8 Melissa is planning to start a business selling ice-cream. She presents a business plan to her bank manager in which she models the situation as follows.

'Three tenths of days are *warm*, two fifths are *normal* and the rest are *cool*. On warm days I will make £100 (that is the difference between what I get from selling ice-cream and the cost of the ingredients), on normal days £70 and on cool days £50. My additional daily costs will be £45.'
  (a) Find Melissa's expected profit per six-day week.
  (b) Find Melissa's expected profit after 12 weeks.

9 The discrete random variable X has probability distribution given by

$$P(X = x) = \frac{(4x + 7)}{68} \quad \text{for } x = 1, 2, 3, 4.$$

  (a) Find (i) $E(X)$ (ii) $E(X^2)$ (iii) $E(X^2 + 5X - 2)$.
  (b) Verify that $E(X^2 + 5X - 2) = E(X^2) + 5E(X) - 2$.

10 A discrete random variable X has a probability distribution as follows:

x	0	1	5	20	w
$P(X = x)$	0.1	0.2	0.3	0.1	0.3

Find the value of w in the two cases
  (a) $E(X) = 10$
  (b) $E(X) = 12$.

11 Jasmine and Robert are playing a game in which they roll two ordinary dice to obtain a total score, X.
  (a) Write down the probability distribution of X.

Halfway through the game their dog eats one of the dice. Robert says 'It's all right. We can play with one die and double its score.' This score is denoted by Y.
  (b) Write down the probability distribution of Y.
  (c) Which of the following are the same and which are different?
    (i) $E(X)$ and $E(Y)$.
    (ii) The range of X and the range of Y.
    (iii) The mode of X and the mode of Y.

12 A woman has five coins in a box: two £1, two 20p and one 10p. She wants the 10p coin to use in a slot machine and takes the coins out at random, one at a time until she comes to the one she wants. She does not replace a coin once she has taken it out of the box. Find
  (a) the expectation of the number of coins she takes out
  (b) the expectation of the amount of money she takes out.
  (In both cases include the final 10p.)

13 An experiment consists of throwing two unbiased dice and recording r, the higher of the two numbers shown. When each die shows the same number, this is taken as the value of r. Complete the entries in the following tables, where $p_1, p_2, \ldots$ are the probabilities that $r = 1, 2, \ldots$.

p_1	p_2	p_3	p_4	p_5	p_6
$\frac{1}{36}$	$\frac{3}{36}$				

Verify that the mean of r is $\frac{161}{36}$.

The experiment, as described above, is done *three* times. Find the probabilities that
  (a) the three values of r are 3, 4, 5 in any order
  (b) the sum of the three values of r is 16.

[MEI]

14 A radio network is launching a new music game (based on one actually broadcast in France). Each contestant is given the title of a song and a list of seven words, exactly three of which occur in the lyric of the song. The contestant is asked to choose the three correct words (the choice being made before listening to the song). Assuming that a particular contestant makes this choice completely by guesswork, find the probabilities that he gets 0, 1, 2, 3 words correct.

Prizes of £1, £3 and £r are to be awarded for 1, 2 and 3 words correct respectively. Find, in terms of r, the expected prize paid to a contestant choosing completely by guesswork. Hence determine the greatest integer value of r for which the expected prize is less than £3.

[MEI]

VARIANCE

The standard deviation of a discrete random variable gives you a measure of the spread of the distribution about the mean or expected value. The variance is simply the square of the standard deviation.

Variance

In Chapter 3 you learned that

$$\text{standard deviation} = \sqrt{\frac{\sum x^2}{n} - \bar{x}^2}$$

$$\text{and variance} = \frac{\sum x^2}{n} - \bar{x}^2.$$

Standard deviation was usually preferred to variance because it is in the same units as those of the data. However, mathematically, it is easier if you now work with the variance. The definition of the variance of a discrete random variable is very similar to that used when finding the variance of a set of numbers.

The variance of a discrete random variable X, $\text{Var}(X)$, is given by

$$\text{Var}(X) = E([X - \mu]^2)$$

Another, and often more convenient, form of this definition is

$$\text{Var}(X) = E(X^2) - \mu^2$$

or $\quad\quad \text{Var}(X) = E(X^2) - [E(X)]^2$

Proof that the two forms are equivalent:

$$\begin{aligned}\text{Var}(X) &= E([X - \mu]^2) \\ &= E(X^2 - 2\mu X + \mu^2) \\ &= E(X^2) - 2\mu E(X) + E(\mu^2) \\ &= E(X^2) - 2\mu^2 + \mu^2 \\ &= E(X^2) - \mu^2\end{aligned}$$

This result is sometimes written in the form

$$\text{Var}(X) = E(X^2) - [E(X)]^2$$
$$= \text{expectation of the square} - \text{the square of the expectation}.$$

EXAMPLE 6.8

The discrete random variable X has the following probability distribution:

x_i	0	1	2	3
$P(X = x_i)$	0.2	0.3	0.4	0.1

Find
(a) $E(X)$
(b) $E(X^2)$
(c) $\text{Var}(X)$ using **(i)** $E(X^2) - \mu^2$ **(ii)** $E([X - \mu]^2)$.

Solution (a) $E(X) = \sum x_i P(X = x_i)$
$= 0 \times 0.2 + 1 \times 0.3 + 2 \times 0.4 + 3 \times 0.1$
$= 1.4$

(b) $E(X^2) = \sum x_i^2 P(X = x_i)$
$= 0 \times 0.2 + 1 \times 0.3 + 4 \times 0.4 + 9 \times 0.1$
$= 2.8$

(c) (i) $Var(X) = E(X^2) - \mu^2$
$= 2.8 - 1.4^2$
$= 0.84$

(ii) $Var(X) = E([X - \mu]^2)$
$= \sum (x_i - \mu)^2 P(X = x_i)$
$= (0 - 1.4)^2 \times 0.2 + (1 - 1.4)^2 \times 0.3 + (2 - 1.4)^2 \times 0.4$
$+ (3 - 1.4)^2 \times 0.1$
$= 0.392 + 0.048 + 0.144 + 0.256$
$= 0.84$

Notice that the two methods of calculating the variance in part **(c)** give the same result, as of course they must.

EXAMPLE 6.9

The random variable X has the following probability distribution:

x	1	2	3	4
$P(X = x_i)$	0.6	0.2	0.1	0.1

Find
(a) $Var(X)$
(b) $Var(7)$
(c) $Var(3X)$
(d) $Var(3X + 7)$.
What general results do answers **(b)** to **(d)** illustrate?

EXERCISE 6C

Solution

(a)

x^2	1	2	3	4
$P(X = x_i)$	0.6	0.2	0.1	0.1

$E(X) = 1 \times 0.6 + 2 \times 0.2 + 3 \times 0.1 + 4 \times 0.1$
$ = 1.7$

$E(X^2) = 1 \times 0.6 + 4 \times 0.2 + 9 \times 0.1 + 16 \times 0.1$
$ = 3.9$

$Var(X) = E(X^2) - [E(X)]^2$
$ = 3.9 - 1.7^2$
$ = 1.01$

(b) $Var(7) = E(7^2) - [E(7)]^2$ *General result*
$ = E(49) - [7]^2$ $Var(c) = 0$ for a constant c.
$ = 49 - 49$ This result is obvious; a constant is
$ = 0$ constant and so can have no spread.

(c) $Var(3X) = E[(3X)^2] - \mu^2$ *General result*
$ = E(9X^2) - [E(3X)]^2$ $Var(aX) = a^2 Var(X)$.
$ = 9E(X^2) - [3E(X)]^2$ Notice that it is a^2 and not a on the
$ = 9 \times 3.9 - (3 \times 1.7)^2$ right-hand side, but that taking
$ = 35.1 - 26.01$ the square root of each side gives
$ = 9.09$ the standard deviation $(aX) =$
   $a \times$ standard deviation (X) as you
   would expect from common sense.

(d) $Var(3X + 7)$ *General result*
$= E[(3X + 7)^2] - [E(3X + 7)]^2$ $Var(aX + c) = a^2 Var(X)$.
$= E(9X^2 + 42X + 49) - [3E(X) + 7]^2$ Notice that the constant c does
$= E(9X^2) + E(42X) + E(49)$ not appear on the right-hand side.
$ - [3 \times 1.7 + 7]^2$
$= 9E(X^2) + 42E(X) - 12.1^2$
$= 9 \times 3.9 + 42 \times 1.7 - 146.41$
$= 9.09$

EXERCISE 6C

1 The probability distribution of a random variable X is as follows:

x	1	2	3	4	5
$P(X = x)$	0.1	0.2	0.3	0.3	0.1

(a) Find (i) $E(X)$ (ii) $Var(X)$.
(b) Verify that $Var(2X) = 4Var(X)$.

2 The probability distribution of a random variable X is as follows:

x	0	1	2
$P(X = x)$	0.5	0.3	0.2

(a) Find (i) $E(X)$ (ii) $Var(X)$.

(b) Verify that $Var(5X + 2) = 25 Var(X)$.

3 Prove that $Var(aX - b) = a^2 Var(X)$ where a and b are constants.

4 A coin is biased so that the probability of obtaining a tail is 0.75. The coin is tossed four times and the random variable X is the number of tails obtained. Find

(a) $E(2X)$

(b) $Var(3X)$.

5 Birds of a particular species lay either 0, 1, 2 or 3 eggs in their nests, with probabilities as shown in the following table.

Number of eggs	0	1	2	3
Probability	0.25	0.35	0.30	k

Find

(a) the value of k

(b) the expected number of eggs laid in a nest

(c) the standard deviation of the number of eggs laid in a nest.

[Cambridge]

6 A committee of three is to be selected at random from three men and four women. The number of men on the committee is the random variable Y. Find

(a) $E(Y)$

(b) $Var(Y)$.

7 A discrete random variable W has the distribution:

w	1	2	3	4	5	6
$P(W = w)$	0.1	0.2	0.1	0.2	0.1	0.3

Find the mean and variance of

(a) $W + 7$

(b) $6W - 5$.

8 The random variable X is the number of heads obtained when four unbiased coins are tossed. Construct the probability distribution for X and find
   (a) $E(X)$
   (b) $Var(X)$
   (c) $Var(3X + 4)$.

9 A shop sells red and blue refills for pens, and keeps them all in the same box. The box contains five red refills at £2.00 each and four blue ones at £1.60 each. A customer takes three refills out at random, not realising there are different colours in the box. Find the expectation and variance of the amount of money the customer spends.

10 A board game is played by moving a counter S squares forwards at a time, where S is determined by the following rule.

   A fair six-sided die is thrown once. S is half the number shown on the die if that number is even; otherwise S is twice the number shown on the die.

   Write out a table showing the possible values of S and their probabilities. Use your table to calculate the mean and variance of S.
   [Cambridge]

11 In an arcade game, whenever a lever is pulled a number appears on a screen. The number can be 1, 2 or 3 and the probabilities of obtaining these numbers are given in the following table.

Number	1	2	3
Probability	0.1	0.5	0.4

   (a) The score, S_1, when the lever is pulled for the first time is the number obtained. Calculate the mean and variance of S_1.
   (b) A second score, S_2, is obtained as follows. If $S_1 = 1$ the lever is pulled a second time, and S_2 is the number which then appears on the screen. If $S_1 = 2$ or $S_1 = 3$, the lever is not pulled a second time and S_2 is 0. Tabulate the possible values of $S_1 + S_2$ and their associated probabilities. Hence calculate the mean of $S_1 + S_2$.
   [Cambridge]

12 An electronic device produces an output of 0, 1 or 3 volts, with probabilities $\frac{1}{2}$, $\frac{1}{3}$ and $\frac{1}{6}$ respectively. The random variable X denotes the result of adding the outputs for two such devices, which act independently.
   (a) Show that $P(X = 4) = \frac{1}{9}$.
   (b) Tabulate all the possible values of X with their corresponding probabilities.
   (c) Hence calculate $E(X)$ and $Var(X)$, giving your answers as fractions in their lowest terms.
   [Cambridge]

13 A box contains nine numbered balls. Three balls are numbered 3, four balls are numbered 4 and two balls are numbered 5.

Each trial of an experiment consists of drawing two balls without replacement and recording the sum of the numbers on them, which is denoted by X. Show that the probability that $X = 10$ is $\frac{1}{36}$, and find the probabilities of all other possible values of X.

Use your results to show that the mean of X is $\frac{70}{9}$, and find the standard deviation of X.

Two trials are made. (The two balls in the first trial are replaced in the box before the second trial.) Find the probability that the second value of X is greater than or equal to the first value of X.

[MEI]

14 A curiously shaped six-faced die produces scores, X, for which the probability distribution is given in the following table.

r	1	2	3	4	5	6
$P(X = r)$	k	$\frac{k}{2}$	$\frac{k}{3}$	$\frac{k}{4}$	$\frac{k}{5}$	$\frac{k}{6}$

Show that the constant k is $\frac{20}{49}$, and find the mean and variance of X.

Show that, when this die is thrown twice, the probability of obtaining two equal scores is very nearly $\frac{1}{4}$.

[MEI]

15 A bag contains four balls, numbered 2, 4, 6, 8 but identical in all other respects. One ball is chosen at random and the number on it is denoted by N, so that $P(N = 2) = P(N = 4) = P(N = 6) = P(N = 8) = \frac{1}{4}$.

Show that $\mu = E(N) = 5$ and $\sigma^2 = \text{Var}(N) = 5$.

Two balls are chosen at random one after the other, with the first ball being replaced after it has been drawn. Let $\overline{N}$ be the arithmetic mean of the numbers on the two balls. List the possible values of $\overline{N}$ and their probabilities of being obtained. Hence evaluate $E(\overline{N})$ and $\text{Var}(\overline{N})$.

[MEI]

16 A random number generator in a computer game produces values which can be modelled by the discrete random variable X with probability distribution given by $P(X = r) = kr!$ for $r = 0, 1, 2, 3, 4$ where k is a constant.

(a) Show that $k = \frac{1}{34}$ and illustrate the probability distribution with a sketch.

(b) Find the expectation and variance of X.

Two independent values of X are generated. Let these values be X_1 and X_2.

(c) Show that $P(X_1 = X_2)$ is a little greater than 0.5.

(d) Given that $X_1 = X_2$, find the probability that X_1 and X_2 are each equal to 4.

[MEI]

Exercise 6C

17 A traffic surveyor is investigating the lengths of queues at a particular set of traffic lights during the daytime, but outside rush hours. He counts the number of cars, X, stopped and waiting when the lights turn green on 90 different occasions, with the following results:

No. of cars, x_i	0	1	2	3	4	5	6	7	8	9+
Frequency, f_i	3	9	12	15	15	15	11	8	2	0

(a) Use these figures to estimate the probability distribution of the number of cars waiting when the lights turn green.

(b) Use your probability distribution to estimate the expectation and variance of X.

A colleague of the surveyor suggests that the probability distribution might be modelled by the expression $P(X = x_i) = kx_i(8 - x_i)$.

(c) Find the value of k.

(d) Find the values of the expectation and variance of X given by this model.

(e) Do you think it is a good model?

18 I was asked recently to analyse the number of goals scored per game by our local Ladies' Hockey Team. Having studied the results for the whole of last season, I proposed the following model, where the discrete random variable X represents the number of goals scored per game by the team:

$$P(X = r) = k(r + 1)(5 - r)^2 \quad \text{for } r = 0, 1, 2, 3.$$

(a) Show that k is 0.01 and illustrate the probability distribution with a sketch.

(b) Find the expectation and variance of X.

(c) Assuming that the model is valid for the forthcoming season, find the probability that

(i) the team will fail to score in the first two games

(ii) the team will score a total of four goals in the first two games.

What other assumption is necessary to obtain these answers?

(d) Give two distinct reasons why the model might not be valid for the forthcoming season.

[MEI]

THE UNIFORM DISTRIBUTION

This distribution occurs when all the possible outcomes are equally likely, as for example when a single die is rolled. There are six possible outcomes, each with the same probability, $\frac{1}{6}$, as shown in figure 6.6.

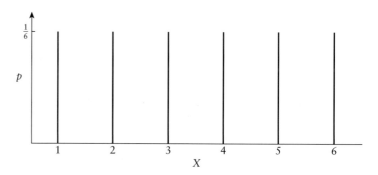

FIGURE 6.6 *Probability distribution of the score when a single die is rolled*

The mean and variance of this distribution are the same as those of the particular values the variable can take.

EXAMPLE 6.10

A fair die is thrown once, and the score X is recorded.
(a) Find $E(X)$ and $Var(X)$.

The random variable Y is defined as $Y = 4X + 3$. Deduce the values of
(b) $E(Y)$ and $Var(Y)$.

Solution **(a)** The distribution of X is

r	1	2	3	4	5	6
$P(X = r)$	$\frac{1}{6}$	$\frac{1}{6}$	$\frac{1}{6}$	$\frac{1}{6}$	$\frac{1}{6}$	$\frac{1}{6}$

$$\therefore E(X) = \sum x_i p_i$$
$$= 1 \times \tfrac{1}{6} + 2 \times \tfrac{1}{6} + 3 \times \tfrac{1}{6} + 4 \times \tfrac{1}{6} + 5 \times \tfrac{1}{6} + 6 \times \tfrac{1}{6}$$
$$= 3\tfrac{1}{2}$$

$$Var(X) = 1^2 \times \tfrac{1}{6} + 2^2 \times \tfrac{1}{6} + 3^2 \times \tfrac{1}{6} + 4^2 \times \tfrac{1}{6} + 5^2 \times \tfrac{1}{6} + 6^2 \times \tfrac{1}{6} - \left[3\tfrac{1}{2}\right]^2$$
$$= \tfrac{91}{6} - \tfrac{49}{4}$$
$$= \tfrac{35}{12}$$
$$= 2\tfrac{11}{12}$$

(b) Since $Y = 4X + 3$

$$E(Y) = 4E(X) + 3$$
$$= 4 \times 3\tfrac{1}{2} + 3$$
$$= 17$$

$$\text{Var}(Y) = 4^2 \text{Var}(X)$$
$$= 16 \times 2\tfrac{11}{12}$$
$$= 46\tfrac{2}{3}$$

Exercise 6D

1 In a toy cricket game the number of runs, X, scored from each ball is equally likely to be 0, 1, 2, 3 or 4.
  (a) Tabulate the distribution of X.
  (b) Calculate $E(X)$ and $\text{Var}(X)$.

  To speed the game up, a boy writes down Y where $Y = 2X + 1$.
  (c) Find $E(Y)$ and $\text{Var}(Y)$.

2 A spinner has four equal sectors, marked with the numbers 1, 2, 3 and 4. The letter X represents the score when the spinner is spun once.
  (a) Name the distribution of X, and write out its distribution.
  (b) Find $E(X)$ and $\text{Var}(X)$.

  The spinner is now spun twice, and the total score $Y = X_1 + X_2$ is recorded.
  (c) Use a sample space diagram to show that the different values of Y are not equally likely.

3 A random number generator produces a single digit number X between 0 and 9 inclusive. Each possible result is equally likely.
  (a) Write out the probability distribution of X.
  (b) Write out the cumulative distribution for X.
  (c) Find $E(X)$ and $\text{Var}(X)$.
  (d) The generator is used to produce a list of 400 numbers. How many of them do you expect to take the value 3 or less?

4 A doctor decides to model the number of cases, X, of flu that she sees per day, as follows:

r	0	1	2	3
$P(X = r)$	p	p	p	p

  (a) Find the value of p.
  (b) Find $E(X)$ and $\text{Var}(X)$.
  (c) Suggest two aspects of the doctor's model which do not resemble real behaviour.

5 Meera is playing a card game. From a normal pack of 52 she chooses a card at random, and notes whether it is a club (3 points), diamond (5 points), heart (7 points) or spade (9 points). She records the number of points, X, replaces the card, then shuffles the pack before drawing another.
  (a) Write out the distribution of X.
  (b) Find E(X) and Var(X).

  Meera's friend Nick plays a similar game, but he records the score Y based on a club (0 points), diamond (1 point), heart (2 points) or a spade (3 points).
  (c) Find E(Y) and Var(Y). You may do this directly, or by adapting your results from (b) above.

EXERCISE 6E Examination style questions

1 A fair six-sided die is rolled. The random variable Y represents the score on the uppermost face.
  (a) Write down the probability function of Y.
  (b) State the name of the distribution of Y.

  Find the value of
  (c) E(6Y + 2)
  (d) Var(4Y − 2).

  [Edexcel]

2 The random variable, X, is the number of sixes obtained when three fair dice are rolled.
  (a) Explain briefly why the distribution of X is not uniform.
  (b) Show that P(X = 3) = 0.0046.
  (c) Copy and complete the table giving the probability distribution of X.

r	0	1	2	3
P(X = r)	0.5787	0.3742		

  (d) Calculate E(X) and Var(X).

3 A discrete random variable X has the probability function shown in the table below.

x	0	1	2
P(X = x)	$\frac{1}{3}$	a	$\frac{2}{3} - a$

  (a) Given that E(X) = $\frac{5}{6}$, find a.
  (b) Find the exact value of Var(X).
  (c) Find the exact value of P(X ⩽ 1.5).

  [Edexcel]

4 The discrete random variable X has the following probability distribution.

x	−2	−1	0	1	2
$P(X = x)$	α	0.2	0.1	0.2	β

(a) Given that $E(X) = -0.2$, find the value of α and the value of β.
(b) Write down $F(0.8)$.
(c) Evaluate $Var(X)$.

Find the value of
(d) $E(3X - 2)$
(e) $Var(2X + 6)$.

[Edexcel]

5 Dan's hobby is archery. When shooting at a target, he scores 1, 2 or 3 points when he hits the target, depending on how close he is to the centre. If he misses the target completely he scores 0 points. From past experience, the distribution of X, his score for each shot, is as follows:

x	0	1	2	3
$P(X = r)$	0.1	0.4	0.3	0.2

(a) Tabulate the cumulative distribution function for X, i.e. $P(X \leqslant r)$ for $r = 0, 1, 2, 3$.

Dan has three shots at the target and his scores are independent. Let L be the discrete random variable which represents the largest of the three scores.
(b) Explain why $P(L \leqslant 2) = [P(X \leqslant 2)]^3 = 0.512$.
Find similarly $P(L \leqslant 1)$.
Write down $P(L = 0)$ and $P(L \leqslant 3)$.
(c) Use the probabilities calculated in part (b) to write down the probability distribution for L.
(d) Find the mean and variance of L.

[MEI]

6 In each round of a general knowledge quiz a contestant is asked up to three questions. The round stops when the contestant gets a question wrong or has answered all three questions correctly. One point is awarded for each correct answer. A contestant who answers all three questions correctly receives in addition one bonus point.

Jayne's probability of getting any particular question correct is 0.7, independently of other questions. Let X represent the number of points she scores in a round.
(a) Show that $P(X = 1) = 0.21$ and find $P(X = r)$ for $r = 0, 2$ and 4.
(b) Draw a sketch to illustrate this discrete probability distribution.
(c) Find the mean and standard deviation of X.
(d) Find the probability that Jayne obtains a higher score in the second round than she does in the first round.

[MEI]

7 The discrete random variable X has the probability function shown in the table below.

x	1	2	3	4	5
$P(X = x)$	0.2	0.3	0.3	0.1	0.1

Find
(a) $P(2 < X \leqslant 4)$
(b) $F(3.7)$
(c) $E(X)$
(d) $Var(X)$
(e) $E(X^2 + 4X - 3)$.

[Edexcel]

8 When a certain type of cell is subjected to radiation, the cell may die, survive as a single cell or divide into two cells, with probabilities $\frac{1}{2}, \frac{1}{3}, \frac{1}{6}$ respectively.

Two cells are independently subjected to radiation. The random variable X represents the total number of cells in existence after this experiment.
(a) Show that $P(X = 2) = \frac{5}{18}$.
(b) Find the probability distribution for X.
(c) Evaluate $E(X)$.
(d) Show that $Var(X) = \frac{10}{9}$.

Another two cells are submitted to radiation in a similar experiment and the random variable Y represents the total number of cells in existence after this experiment. The random variable Z is defined as $Z = X - Y$.
(e) Find $E(Z)$ and $Var(Z)$.

[Edexcel]

9 The discrete random variables A and B are independent and have the following probability distributions:

a	1	2	3
$P(A = a)$	$\frac{1}{4}$	$\frac{1}{2}$	$\frac{1}{4}$

b	1	2
$P(B = b)$	$\frac{1}{3}$	$\frac{2}{3}$

The random variable Q is the product of one observation from A and one observation from B.

(a) Show that $P(Q = 2) = \frac{1}{3}$.
(b) Find the probability distribution for Q.
(c) Hence, or otherwise, show that $E(Q) = \frac{10}{3}$.
(d) Find $\text{Var}(Q)$.

[Edexcel]

10 An economics student is trying to model the daily movement, X points, in a stock market indicator. The student assumes that the value of X on one day is independent of the value on the next day. A fair die is rolled and if an odd number is uppermost then the indicator is moved down that number of points. If an even number is uppermost then the indicator is moved up that number of points.

(a) Write down the distribution of X as specified by the student's model.
(b) Find the value of $E(X)$.
(c) Show that $\text{Var}(X) = \frac{179}{12}$.

If the indicator moves upwards over a period of time then this is taken as a sign of growth in the economy, if it falls then this is a sign that the economy is in decline.

(d) Comment on the state of the economy as suggested by this model.

Before the stockmarket opened one Monday morning the economic indicator was 3373.

(e) Use the student's model to find the probability that the indicator is at least 3400 when the stockmarket closes on the Friday afternoon of the same week.

[Edexcel, part]

KEY POINTS

1. For a discrete random variable, X, which takes values $x_1, x_2, x_3, \ldots$ with probabilities $p_1, p_2, p_3, \ldots$

   - $\sum p_i = 1$ where $0 \leq p_i \leq 1$
   - $E(X) = \sum x_i p_i$
   - $Var(X) = E(X^2) - [E(X)]^2$

2. The cumulative distribution function $F(x)$ is given by

   - $F(x_0) = P(X \leq x_0) = \sum_{x \leq x_0} p_x$

3. For any discrete random variable X and constants a and b:

   - $E(b) = b$
   - $E(aX) = aE(X)$
   - $E(aX + b) = aE(X) + b$
   - $Var(b) = 0$
   - $Var(aX) = a^2 Var(X)$
   - $Var(aX + b) = a^2 Var(X)$

4. If a random variable X is such that the values $x_1, x_2, x_3, \ldots$ occur with equal probability then X is said to follow a discrete uniform distribution.

Chapter seven

The normal distribution

The
normal
law of error
stands out in the
experience of mankind
as one of the broadest
generalisations of natural
philosophy. It serves as the
guiding instrument in researches
in the physical and social sciences
and in medicine, agriculture and engineering.
It is an indispensable tool for the analysis and the
interpretation of the basic data obtained by observation and experiment.

W. J. Youden

Wilf Roberts enjoys playing basketball. At 195 cm his height helps him play the game to a high standard. Wilf is clearly exceptionally tall, but how much so? Is he one in a hundred, or a thousand or even a million? To answer that question you need to know the distribution of the heights of men.

The first point that needs to be made is that height is a continuous variable and not a discrete one. If you measure accurately enough it can take any value.

This means that it does not really make sense to ask 'What is the probability that somebody chosen at random has height exactly 195 cm?' The answer is zero.

However, you can ask questions like 'What is the probability that somebody chosen at random has height between 194 cm and 196 cm?' and 'What is the probability that somebody chosen at random has height at least 195 cm?'. When the variable is continuous, you are concerned with a range of values rather than a single value.

Like many other naturally occurring variables, the heights of men may be modelled by the normal distribution, shown below. You will see that this has a distinctive bell-shaped curve and is symmetrical about its middle. The curve is continuous as height is a continuous variable.

FIGURE 7.1

On figure 7.1, area represents probability so the shaded area to the right of 195 cm represents the probability that a randomly selected adult male is over 195 cm tall.

Before you can start to find this area, you must know the mean and standard deviation of the distribution, in this case about 174 cm and 7 cm respectively.

So Wilf's height is 195 cm − 174 cm = 21 cm above the mean, and that is $\frac{21}{7}$ = 3 standard deviations. The number of standard deviations beyond the mean, in this case 3, is denoted by the letter z. Thus the shaded area gives the probability of obtaining a value of $z \geqslant 3$.

You find this area by looking up the value of $\Phi(z)$ when $z = 3$ in a normal distribution table of $\Phi(z)$ as shown in figure 7.2, and then calculating $1 - \Phi(z)$. (Φ is the Greek letter phi.)

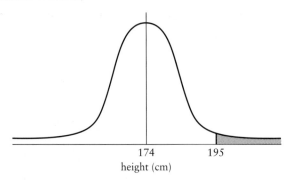

z	$\Phi(z)$	z	$\Phi(z)$	z	$\Phi(z)$	z	$\Phi(z)$	z	$\Phi(z)$
0.29	0.6141	0.79	0.7852	1.29	0.9015	1.79	0.9633	2.70	0.9965
0.30	0.6179	0.80	0.7881	1.30	0.9032	1.80	0.9641	2.75	0.9970
0.31	0.6217	0.81	0.7910	1.31	0.9049	1.81	0.9649	2.80	0.9974
0.32	0.6255	0.82	0.7939	1.32	0.9066	1.82	0.9656	2.85	0.9978
0.33	0.6293	0.83	0.7967	1.33	0.9082	1.83	0.9664	2.90	0.9981
0.34	0.6331	0.84	0.7995	1.34	0.9099	1.84	0.9671	2.95	0.9984
0.35	0.6368	0.85	0.8023	1.35	0.9115	1.85	0.9678	3.00	0.9987
0.36	0.6406	0.86	0.8051	1.36	0.8131	1.86	0.9686	3.05	0.9989
0.37	0.6443	0.87	0.8078	1.37	0.9147	1.87	0.9693	3.10	0.9990

FIGURE 7.2 *Extract from tables of $\Phi(z)$*

This gives $\Phi(3) = 0.9987$, and so $1 - \Phi(3) = 0.0013$.

The probability of a randomly selected adult male being 195 cm or over is 0.0013. Slightly more than one man in a thousand is at least as tall as Wilf.

Using normal distribution tables

The function $\Phi(z)$ gives the area under the normal distribution curve to the *left* of the value z, that is the shaded area in figure 7.3 (it is the cumulative distribution function). The total area under the curve is 1, and the area given by $\Phi(z)$ represents the probability of a value smaller than z.

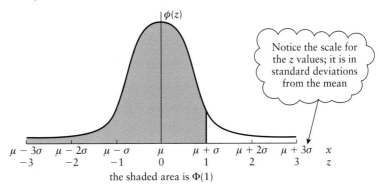

Figure 7.3

the shaded area is $\Phi(1)$

If the variable X has mean μ and standard deviation σ then x, a particular value of X, is transformed into z by the equation

$$z = \frac{x - \mu}{\sigma}$$

z is a particular value of the variable Z which has mean 0 and standard deviation 1 and is the *standardised* form of the normal distribution.

	Actual distribution, X	Standardised distribution, Z
Mean	μ	0
Standard deviation	σ	1
Particular value	x	$z = \dfrac{x - \mu}{\sigma}$

Notice how lower case letters, x and z, are used to indicate particular values of the random variables, whereas upper case letters, X and Z, are used to describe or name those variables.

Normal distribution tables are easy to use but you should always make a point of drawing a diagram and shading the region you are interested in.

It is often helpful to know that in a normal distribution, roughly:

- 68% of the values lie within ±1 standard deviation of the mean;
- 95% of the values lie within ±2 standard deviations of the mean;
- 99.75% of the values lie within ±3 standard deviations of the mean.

7 THE NORMAL DISTRIBUTION

EXAMPLE 7.1

Assuming the distribution of the heights of adult men is normal, with mean 174 cm and standard deviation 5 cm, find the probability that a randomly selected adult man is

(a) under 185 cm
(b) over 185 cm
(c) over 180 cm
(d) between 180 cm and 185 cm
(e) under 170 cm

giving answers to 2 significant figures.

Solution The mean height, $\mu = 174$
The standard deviation, $\sigma = 5$

(a) The probability that an adult man selected at random is under 185 cm.

The area required is that shaded in the diagram.

$x = 185$ cm

and so $z = \dfrac{185 - 174}{5} = 2.2$

$\Phi(2.2) = 0.9861$
$= 0.98$ (2 sf)

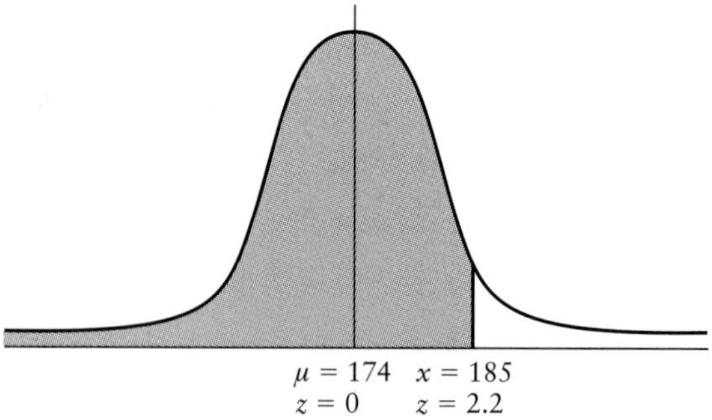

FIGURE 7.4

$\mu = 174 \quad x = 185$
$z = 0 \quad\quad z = 2.2$

Answer: The probability that an adult man selected at random is under 185 cm is 0.98.

(b) The probability that an adult man selected at random is over 185 cm.

The area required is the complement of that for part (a).

Probability $= 1 - \Phi(2.2)$
$= 1 - 0.9861$
$= 0.0139$
$= 0.014$ (2 sf)

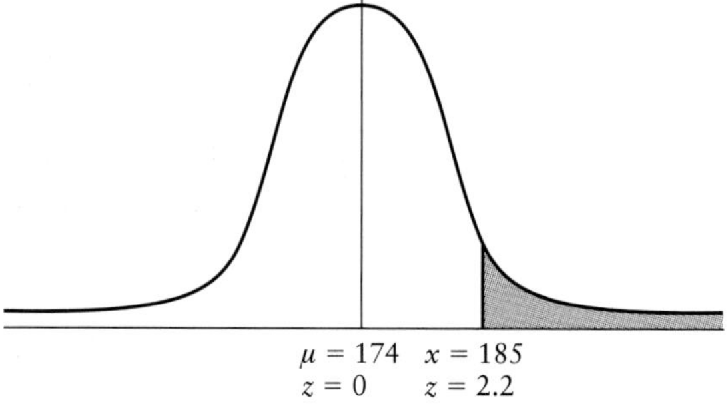

FIGURE 7.5

$\mu = 174 \quad x = 185$
$z = 0 \quad\quad z = 2.2$

Answer: The probability that an adult man selected at random is over 185 cm is 0.014.

(c) The probability that an adult man selected at random is over 180 cm.

$x = 180$ and so $z = \dfrac{180 - 174}{5} = 1.2$

The area required = $1 - \Phi(1.2)$
= $1 - 0.8849$
= 0.1151
= 0.12 (2 sf)

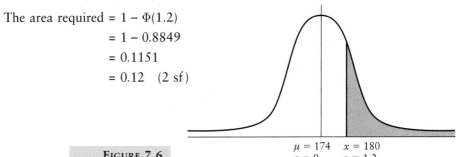

FIGURE 7.6

Answer: The probability that an adult man selected at random is over 180 cm is 0.12.

(d) The probability that an adult man selected at random is between 180 cm and 185 cm.

The required area is shown in the diagram. It is

$\Phi(2.2) - \Phi(1.2) = 0.9861 - 0.8849$
= 0.1012
= 0.10 (2 sf)

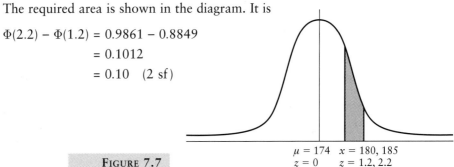

FIGURE 7.7

Answer: The probability that an adult man selected at random is over 180 cm but under 185 cm is 0.10.

(e) The probability that an adult man selected at random is under 170 cm.

In this case $x = 170$

and so $z = \dfrac{170 - 174}{5} = -0.8$

FIGURE 7.8

However when you come to look up Φ(−0.8), you will find that only positive values of z are given in your tables. You overcome this problem by using the symmetry of the normal curve. The area you want in this case is that to the left of −0.8 and this is clearly just the same as that to the right of +0.8.

So Φ(−0.8) = 1 − Φ(0.8)
= 1 − 0.7881 = 0.2119
= 0.21   (2 sf)

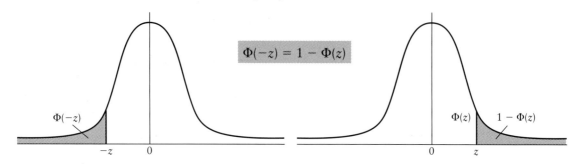

Φ(−z) = 1 − Φ(z)

FIGURE 7.9

Answer: The probability that an adult man selected at random is under 170 cm is 0.21.

THE NORMAL CURVE

All normal curves have the same basic shape, so that by scaling the two axes suitably you can always fit one normal curve exactly on top of another one.

The curve for the normal distribution with mean μ and standard deviation σ (i.e. variance σ^2) is given by the function $\phi(x)$ in

$$\phi(x) = \frac{1}{\sigma\sqrt{2\pi}} e^{-\frac{1}{2}\left(\frac{x-\mu}{\sigma}\right)^2}$$

The notation $N(\mu, \sigma^2)$ is used to describe this distribution. The mean, μ, and standard deviation, σ (or variance, σ^2), are the two parameters used to define the distribution. Once you know their values, you know everything there is to know about the distribution. The standardised variable Z has mean 0 and variance 1, so its distribution is $N(0, 1)$.

After the variable X has been transformed to Z using $z = \frac{x-\mu}{\sigma}$ the form of the curve (now standardised) becomes

$$\phi(z) = \frac{1}{\sqrt{2\pi}} e^{-\frac{1}{2}z^2}$$

However, the exact shape of the normal curve is often less useful than the area underneath it, which represents a probability. For example, the probability that $Z \leq 2$ is given by the shaded area in figure 7.10.

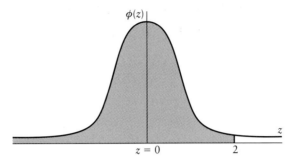

FIGURE 7.10

Easy though it looks, the function $\phi(z)$ cannot be integrated algebraically to find the area under the curve, only by a numerical method. The values found by doing so are given as a table and this area function is called $\Phi(z)$.

EXAMPLE 7.2

Skilled operators make a particular component for an engine. The company believes that the time taken to make this component may be modelled by the normal distribution with mean 95 minutes and standard deviation 4 minutes.

Assuming the company's belief to be true find the probability that the time taken to make one of these components, selected at random, was

(a) over 97 minutes
(b) under 90 minutes
(c) between 90 and 97 minutes.

Solution According to the company $\mu = 95$ and $\sigma = 4$ so the distribution is $N(95, 4^2)$.

(a) The probability that a component required over 97 minutes.

$$z = \frac{97 - 95}{4} = 0.5$$

The probability is represented by the shaded area and given by

$1 - \Phi(0.5) = 1 - 0.6915$
$= 0.3085$
$= 0.31 \quad (2 \text{ dp})$

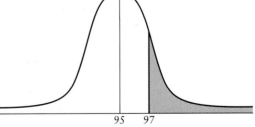

FIGURE 7.11

Answer: The probability it took the operator over 97 minutes to manufacture a randomly selected component is 0.31.

(b) The probability that a component required under 90 minutes.

$$z = \frac{90 - 95}{4} = -1.25$$

The probability is represented by the shaded area and given by

$1 - \Phi(1.25) = 1 - 0.8944$
$ = 0.1056$
$ = 0.11 \quad (2\text{ dp})$

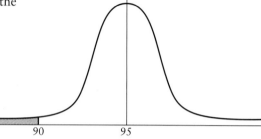

FIGURE 7.12

Answer: The probability it took the operator under 90 minutes to manufacture a randomly selected component is 0.11.

(c) The probability that a component required between 90 and 97 minutes.

The probability is represented by the shaded area and given by

$1 - 0.1056 - 0.3085 = 0.5859$
$ = 0.59 \quad (2\text{ dp})$

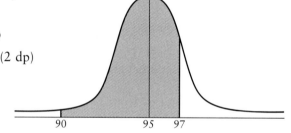

FIGURE 7.13

Answer: The probability it took the operator between 90 and 97 minutes to manufacture a randomly selected component is 0.59.

EXERCISE 7A

1 The distribution of the heights of 18-year-old girls may be modelled by the normal distribution with mean 162.5 cm and standard deviation 6 cm. Find the probability that the height of a randomly selected 18-year-old girl is
  (a) under 168.5 cm
  (b) over 174.5 cm
  (c) between 168.5 and 174.5 cm.

2 A pet shop has a tank of goldfish for sale. All the fish in the tank were hatched at the same time and their weights may be taken to be normally distributed with mean 100 g and standard deviation 10 g. Melanie is buying a goldfish and is invited to catch the one she wants in a small net. In fact the fish are much too quick for her to be able to catch any particular one and the fish which she eventually nets is selected at random. Find the probability that its weight is
  (a) over 115 g   (b) under 105 g   (c) between 105 and 115 g.

EXERCISE 7A

3 When he makes instant coffee, Tony puts a spoonful of powder into a mug. The weight of coffee in grams on the spoon may be modelled by the normal distribution with mean 5 and standard deviation 1. If he uses more than 6.5 g Julia complains that it is too strong and if he uses less than 4 g she tells him it is too weak. Find the probability that he makes the coffee
   (a) too strong   (b) too weak   (c) all right.

4 When a butcher takes an order for a Christmas turkey, he asks the customer what weight in kilograms the bird should be. He then sends his order to a turkey farmer who supplies birds of about the requested weight. For any particular weight of bird ordered, the error in kilograms may be taken to be normally distributed with mean 0 and standard deviation 0.8.

   Mrs Jones orders a 10 kg turkey from the butcher. Find the probability that the one she gets is
   (a) over 12 kg
   (b) under 10 kg
   (c) within 0.6 kg of the weight she actually ordered.

5 A normally distributed random variable X has mean 20.0 and variance 6.25. Find the probability that $18.0 < X < 21.0$.

6 The concentration by volume of methane at a point on the centre line of a jet of natural gas mixing with air is distributed approximately normally with mean 20% and standard deviation 10%. Find the probabilities that the concentration
   (a) exceeds 30%
   (b) is between 5% and 15%.
   (c) In another similar jet, the mean concentration is 18% and the standard deviation is 5%. Find the probability that in at least one of the jets the concentration is between 5% and 15%.
   [MEI, adapted]

7 A firm makes two different brands of battery, A and B. The life of brand A has mean 23 hours, standard deviation 2 hours; the life of brand B has mean 25 hours, standard deviation 5 hours. Their lives are assumed to be distributed according to a normal probability model.
   Which brand is more likely to stop working over a period of
   (a) 22 hours   (b) 20 hours?

8 A machine produces crankshafts whose diameters are normally distributed with mean 5 cm and standard deviation 0.03 cm. Find the percentage of crankshafts it will produce whose diameters lie between 4.94 cm and 4.97 cm.

   What is the probability that two successive crankshafts will both have a diameter in this interval?
   [MEI, adapted]

Using normal tables in reverse

So far you have been working with normal distributions whose mean and standard deviation are known. Sometimes, however, their values have to be inferred from probabilities which are given. This requires using the normal distribution in reverse, i.e. *percentage points of the normal distribution*.

EXAMPLE 7.3

It is believed that the mean mass of a certain species of reptile is 65 g, and that 20% of them are over 70 g.

Assuming these figures to be reliable observations from a normal distribution, obtain a value for the standard deviation of the mass in grams, correct to 1 decimal place.

Solution Let X represent the mass of a randomly chosen reptile of this species.

Then $X \sim N(65, \sigma^2)$ where σ is to be found.

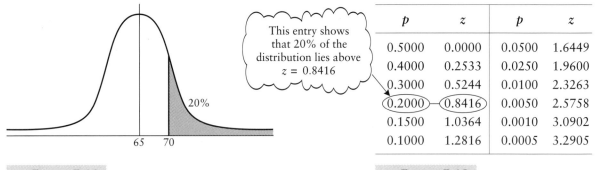

FIGURE 7.14   FIGURE 7.15

From tables of the percentage points of the normal distribution, the z-value corresponding to a 20% upper tail is 0.8416.

$$\therefore \quad \Phi\left(\frac{70-65}{\sigma}\right) = 0.20$$

$$\frac{70-65}{\sigma} = 0.8416$$

$$\frac{5}{\sigma} = 0.8416$$

$$\sigma = \frac{5}{0.8416}$$

$$= 5.941$$

Answer: The required standard deviation is 5.9 g (1 dp).

EXAMPLE 7.4

Sheila is working in a factory. She makes components for an engine. The company reckons that the time taken to make each component can be modelled by a normal distribution.

Sheila invites the company to time her at work. They find that only 10% of the components take her over 90 minutes to make, and that 20% take less than 70 minutes.
(a) Use these figures to obtain estimates of the mean and standard deviation of the time that Sheila takes.
(b) Explain why your answers are only estimates.

Solution (a) Let $X \sim N(\mu, \sigma^2)$.

From tables,
$$p = 0.1000 \Rightarrow z = 1.2816$$

$$\therefore \quad \frac{90 - \mu}{\sigma} = 1.2816$$

so $\quad 90 - \mu = 1.2816\sigma$

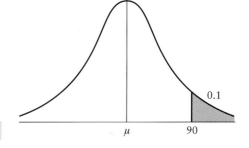

FIGURE 7.16

From tables,
$$p = 0.2000 \Rightarrow z = 0.8416$$

but since we are now at the left (lower) tail we should use $z = -0.8416$

$$\therefore \quad \frac{70 - \mu}{\sigma} = -0.8416$$

$$70 - \mu = -0.8416\sigma$$

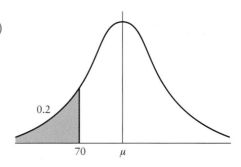

FIGURE 7.17

This gives us two simultaneous equations:
$$90 - \mu = 1.2816\sigma$$
$$70 - \mu = -0.8416\sigma$$

Subtract: $\quad 20 = 2.1232\sigma$

giving $\quad \sigma = 9.41\ldots$ rounding to 9.4
and $\quad \mu = 90 - 1.2816\sigma$
$ = 77.92\ldots$ rounding to 77.9

Answer: Sheila's mean time is 77.9 minutes with standard deviation 9.4 minutes.
(b) These answers are only estimates because they are deduced from a sample of timings, not the full population.

Exercise 7B

1. In a particular experiment, the length of a metal bar is measured many times. The measured values are distributed approximately normally with mean 1.340 m and standard deviation 0.020 m. Find the probabilities that any one measured value
   (a) exceeds 1.370 m
   (b) lies between 1.310 m and 1.370 m
   (c) lies between 1.330 m and 1.390 m.
   (d) Find the length l for which the probability that any one measured value is less than l is 0.1.

   [MEI, adapted]

2. A machine is used to fill cans of soup with a nominal volume of 0.450 litres. Suppose that the machine delivers a quantity of soup which is normally distributed with mean μ litres and standard deviation σ litres. Given that $\mu = 0.457$ and $\sigma = 0.004$, find the probability that a randomly chosen can contains less than the nominal volume.

   It is required by law that no more than 1% of cans contain less than the nominal volume. Find
   (a) the least value of μ which will comply with the law when $\sigma = 0.004$
   (b) the greatest value of σ which will comply with the law when $\mu = 0.457$.

   [MEI]

3. The length of life of a certain make of tyre is normally distributed about a mean of 24 000 km with a standard deviation of 2500 km.
   (a) What percentage of such tyres will need replacing before they have travelled 20 000 km?
   (b) As a result of improvements in manufacture, the length of life is still normally distributed, but the proportion of tyres failing before 20 000 km is reduced to 2.5%.
      (i) If the standard deviation has remained unchanged, calculate the new mean length of life.
      (ii) If, instead, the mean length of life has remained unchanged, calculate the new standard deviation.

   [MEI, adapted]

4. A machine is set to produce nails of lengths 10 cm, with standard deviation 0.05 cm. The lengths of the nails are normally distributed.
   (a) Find the percentage of nails produced between 9.95 cm and 10.08 cm in length.

   The machine's setting is moved by a careless apprentice with the consequence that 15% of the nails are under 5.2 cm in length and 20% are over 5.3 cm.
   (b) Find the new mean and standard deviation.

EXERCISE 7B

5 A factory produces a very large number of steel bars. The lengths of these bars are normally distributed with 30% of them measuring 20.06 cm or more and 10% of them measuring 20.02 cm or less.

Write down two simultaneous equations for the mean and standard deviation of the distribution and solve to find values to 4 significant figures. Hence estimate the proportion of steel bars which measure 20.03 cm or more.

The bars are acceptable if they measure between 20.02 cm and 20.08 cm. What percentage are rejected as being outside the acceptable range?

[MEI]

6 An aircraft component has a life before failure whose mean is 20 months and standard deviation 4 months.
   (a) The component is meant to be replaced on an aircraft so that the probability of it failing in service is 0.1 or less. After how many months should one of these components be replaced?
   (b) One of these components is not replaced when it should be. Calculate the probability that it will last 26 months or more before failure.
   (c) The component is now redesigned so that the standard deviation of its life is changed (but the mean kept the same). The component need now only be replaced after 18 months. Calculate the new standard deviation.

[MEI]

7 The weights of eggs, measured in grams, can be modelled by a $N(85.0, 5^2)$ distribution. Eggs are classified as large, medium or small, where a large egg weighs 90.0 grams or more, and 30% of eggs are classified as small. Calculate
   (a) the percentage of eggs which are classified as large
   (b) the maximum weight of a small egg.

[Cambridge, adapted]

8 Each weekday a man goes to work by bus. His arrival time at the bus stop is normally distributed with standard deviation 4 minutes. His mean arrival time is 8.30 am. Buses leave promptly every 5 minutes at 8.21 am, 8.26 am, etc. Find the probabilities that he catches the buses at
   (a) 8.26 am   (b) 8.31 am   (c) 8.36 am
assuming that he always gets on the first bus to arrive.
   (d) The man is late for work if he catches a bus after 8.31 am. What mean arrival time would ensure that, on average, he is not late for work more than one day in five? [Assume that he cannot change the standard deviation of his arrival time and give your answer to the nearest 10 s.]

[MEI, adapted]

9 A factory is lit by a large number of electric light bulbs whose lifetimes are modelled by a normal distribution with mean 1000 hours and standard deviation 100 hours. Operating conditions require that all bulbs are on continuously.
   (a) What proportions of bulbs have lifetimes that
      (i) exceed 950 hours   (ii) exceed 1050 hours?
   (b) Given that a bulb has already lasted 950 hours, what is the probability that it will last a further 100 hours?
   Give all answers correct to 3 decimal places.

   The factory management is to adopt a policy whereby all bulbs will be replaced periodically after a fixed interval.
   (c) To the nearest day, how long should this interval be if, on average, 1% of the bulbs are to burn out between successive replacement times?

   [MEI, adapted]

10 The mean mark in a mathematics test was 60.
   (a) 40% of the candidates scored 55 or less. Using this information, and the mean given above, obtain a value for the standard deviation of the marks. Assume the marks to be normally distributed.
   (b) 15% of the candidates scored 70 or more. Using this information, and the mean given above, obtain a value for the standard deviation of the marks. Again, assume the marks to be normally distributed.
   (c) Compare your values from (a) and (b). What does this suggest about the distribution of the marks?

EXERCISE 7C   **Examination style questions**

1 The weights of a certain type of apple may be modelled by a normal distribution with mean of 105 g with standard deviation 20 g. One apple is chosen at random.
   (a) Find the probability that its weight is less than 110 g.
   (b) Find the probability that its weight lies between 95 g and 120 g.
   (c) Find the percentage of apples whose weights are over 85 g.

2 Each month I receive a credit card statement, which must be paid by a certain date. Sometimes I send the payment early, while other times it is late. The quantity T denotes the number of days by which a given payment is late. From past records I reckon that T may be modelled by the normal distribution $N(-4, 5^2)$.
   (a) Explain in simple terms the significance of the *minus* sign in this model.
   (b) Calculate the percentage of payments which are at least three days late.
   (c) Calculate the probability that, of my next two payments, one is early and the other is late.

EXERCISE 7C

3 An astrophysicist is studying a star which undergoes outbursts at intervals of a few months. He reckons that the time T days between one outburst and the next can be modelled by the normal distribution $N(70, 8^2)$.
  (a) Find the percentage of outbursts which occur at intervals of less than 80 days.
  (b) Find the probability that two successive outbursts occur at intervals of more than 64 days each.
  (c) Another astrophysicist discovers that long and short intervals tend to alternate. To what extent does your answer to part (b) remain valid in the light of this discovery? Explain your reasoning carefully.

4 Electronic sensors of a certain type fail when they become too hot. The temperature at which a randomly chosen sensor fails is $T\,°C$, where T is modelled as a normal random variable with mean 94.5 and standard deviation 5.5.
  (a) Determine what proportion of sensors will operate in boiling water (i.e. at $100\,°C$).
  (b) The manufacturers wish to quote a safe operating temperature at which 99% of the sensors will work. What temperature should they quote?

5 The continuous random variable Y is normally distributed with mean 100 and variance 256.
  (a) Find $P(Y < 80)$.
  (b) Find k such that $P(100 - k \leqslant Y \leqslant 100 + k) = 0.516$.

  [Edexcel]

6 The random variable X is normally distributed with mean μ and variance σ^2.

  Given that $P(X > 73.8) = 0.025$ and $P(X < 51.335) = 0.4$
  (a) write down two equations containing μ and σ
  (b) solve your equations to find the values of μ and σ.

7 In a reading test for eight-year-old children, it is found that a reading score X is normally distributed with mean 5.0 and standard deviation 2.0.
  (a) What proportion of pupils would you expect to score between 4.5 and 6.0?
  (b) There are about 700 000 eight-year-olds in the country. How many would you expect to have a reading score of more than twice the mean?
  (c) Why might educationalists refer to the reading score X as a 'score out of 10'?

The reading score is often reported, after scaling, as a value Y which is normally distributed, with mean 100 and standard deviation 15. Values of Y are usually given to the nearest integer.

(d) What range of Y scores would you expect to be attained by the best 20% of readers?

[MEI]

8 At a play centre parents pay a fixed fee and may leave their children for as long as they wish. The management's records show that the most common length of stay is 80 minutes and that 25% of the children stay longer than 90 minutes. The length of time a child stays appears to be reasonably well modelled by a normal distribution.

(a) Show the information given on a sketch of the normal distribution. Determine the mean and the standard deviation of the distribution.

[You may assume that, for the standardised normal variable $Z \sim N(0, 1)$, $P(Z > 0.6745) = 0.25$.]

In the rest of the question assume that the distribution is $N(80, 15^2)$.

(b) Calculate the probability that a child stays more than two hours.

The management decide to introduce a closing time of 5 pm.

(c) Explain why the proposed model could not now apply to children arriving at 4 pm.

(d) Give a latest time of arrival for which you consider the model still to be reasonable. Justify your answer.

[MEI]

9 A soft drinks dispenser delivers lemonade into a cup when a coin is inserted into the machine. The amount of lemonade delivered is normally distributed with mean 260 ml and standard deviation 10 ml. The nominal amount of lemonade in a cup is 250 ml. The capacity of the cup is 275 ml.

(a) What is the probability that the cup overflows?
(b) What is the probability that the amount of lemonade in the cup is at least 250 ml but does not overflow?
(c) On one occasion my friends and I purchase five such cups of lemonade. What is the probability that not more than one cup contains less than 250 ml?
(d) Some customers have complained that the proportion of cups giving short measure is too high. The standard deviation of the amount of lemonade delivered per cup is fixed, but the mean can be altered. What value would you recommend so that no more than 5% of cups contain less than 250 ml of lemonade?

[MEI]

Exercise 7C

10 Strips of metal are cut to length L cm, where $L \sim N(\mu, 0.5^2)$.

    (a) Given that 2.5% of the cut lengths exceed 50.98 cm, show that $\mu = 50$.

    (b) Find $P(49.25 < L < 50.75)$.

Those strips with length either less than 49.25 cm or greater than 50.75 cm cannot be used.

Two strips of metal are selected at random.

    (c) Find the probability that both strips cannot be used.

[Edexcel]

KEY POINTS

1. The normal distribution with mean μ and standard deviation σ is denoted by $N(\mu, \sigma^2)$.

2. This may be given in standardised form by using the transformation
$$z = \frac{x - \mu}{\sigma}$$

3. In the standardised form, $N(0, 1)$, the mean is 0, and the standard deviation and variance both 1.

4. The area to the left of the value z in the figure below, representing the probability of a value less than z, is denoted by $\Phi(z)$ and is read from tables.

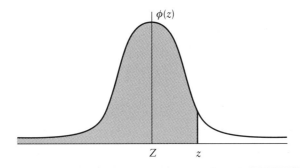

Answers

Chapter 1

Exercise 1A (Page 3)

1. Yes — 0 or 1 equally likely
2. No — 0, 1, 2 not equally likely
3. No — e.g. 12 less likely than 7
4. Yes
5. Yes — (to a fair approximation)
6. No — 0 or 10 are probably less likely than 5 or 6
7. No — 50, 51 more likely than 52, 53
8. No — 0 and 1 more likely than 10 and 11
9. No — Small numbers probably more likely than large numbers
10. No — Again, smaller values are probably more likely than larger ones
11. Yes
12. No — Skewed
13. Possibly yes
14. Yes
15. Yes
16. Yes
17. No — It is possible to be e.g. 10 minutes longer than average, but 10 minutes shorter than average may well be negative!
18. No — Skewed
19. Yes
20. Yes

Chapter 2

Exercise 2A (Page 13)

1. 3.27, 3.32, 3.36, 3.43, 3.45, 3.49, 3.50, 3.52, 3.56, 3.56, 3.58, 3.61, 3.61, 3.64, 3.72

2. 0.083, 0.086, 0.087, 0.090, 0.091, 0.094, 0.098, 0.102, 0.103, 0.105, 0.108, 0.109, 0.109, 0.110, 0.111, 0.114, 0.123, 0.125, 0.131

3. $n = 13$

   21 | 2 represents 0.212

21	2
22	3 6
23	0 3 7
24	1 2 8
25	3 3 9
26	2

4. $n = 10$

   780 | 1 represents 78.01

780	1
790	4 6
800	4 8
810	3 7 9
820	0 5

5. 0.013, 0.089, 1.79, 3.43, 3.51, 3.57, 3.59, 3.60, 3.64, 3.66, 3.68, 3.71, 3.71, 3.73, 3.78, 3.79, 3.80, 3.85, 3.94, 7.42, 10.87

6. (a) $n = 40$

   2 | 8 represents 28 years of age

0	5
1	9
2	8 8 2 6 6 9 9
3	8 7 4 5 7 8 8 6 3 7 9 5 5 2 6 9 3
4	4 6 5 5 1
5	2 9 5
6	2 0 6 1 3
7	
8	1

   (b) Two outliers, 5 years, which is obviously a mistake, and 81 years, which is possible.

7 (a) 16 years of age is a (low) outlier but (with parental permission) it is possible that somebody that age got married. 83 years of age is unusual but not unknown.

(b) 83 years

(c)
```
1 | 6 8 9 9
2 | 0 1 1 1 2 3 4 5 6 6 7 8 8 9
3 | 0 0 0 1 2 2 3 5 9
4 | 3 3 4 5 5 6 6 7 8 9
5 | 1 2 2 7
6 |
7 |
8 | 3
```

(d)
```
1* |
1  | 6 8 9 9
2* | 0 1 1 1 2 3 4
2  | 5 6 6 7 8 8 9
3* | 0 0 0 1 2 2 3
3  | 5 9
4* | 3 3 4
4  | 5 5 6 6 7 8 9
5* | 1 2 2
5  | 7
6* |
6  |
7* |
7  |
8* | 3
```

(e) The stem and leaf diagram with steps of 10 suggests a slight positive skew.

The stretched stem and leaf diagram shows a clear bimodal spread to the distribution. The first peak (20s) may indicate first marriages and the second peak (40s) may indicate second marriages.

Exercise 2B (Page 17)

1 (a) (i) numerical data/discrete
  (ii) a vertical line graph
(b) (i) categorical
  (ii) bar chart
(c) (i) numerical data/continuous
  (ii) histogram or stem and leaf diagram
(d) (i) categorical data
  (ii) bar chart
(e) (i) numerical data/discrete
  (ii) vertical line graph
(f) (i) numerical data/continuous
  (ii) histogram or stem and leaf diagram
(g) (i) categorical data
  (ii) bar chart
(h) (i) numerical data/discrete
  (ii) vertical line graph or bar chart
(i) (i) numerical data/continuous
  (ii) histogram or stem and leaf diagram
(j) (i) categorical data
  (ii) bar chart

2 Seanna: wool £36 000, kelp £24 000, fish £12 000.
Rhos Skerry: wool £45 000, kelp £81 000, fish £36 000.

3 (a)

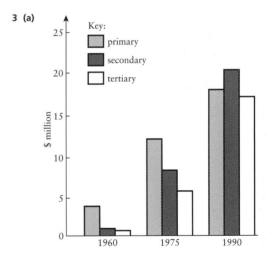

(b)

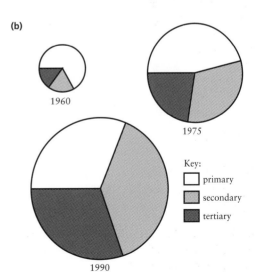

(c) The education budget has increased over the years and proportionally more has been spent on secondary education until in 1990 it overtakes that spent on primary.

(d) The bar chart. It shows the relative amounts spent and indicates the actual amounts spent.

4 (a)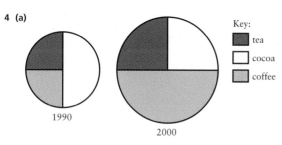

Key:
- tea
- cocoa
- coffee

(b) Crop production has doubled in the period with coffee contributing 50% of all production in 2000 compared with 25% in 1990.

(c) The bar chart because both the relative and actual values are communicated.

5 (a)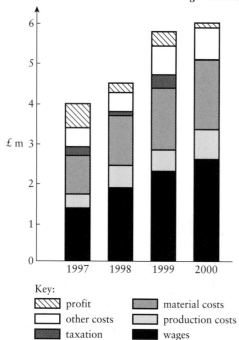

(b) Sales have increased during the period. However, wages have also increased and profits have decreased.

6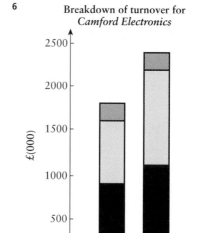

Key:
- profit
- wages
- material costs

EXERCISE 2C (Page 25)

1 (a) $0.5 \leqslant d < 10.5$, $10.5 \leqslant d < 15.5$, $15.5 \leqslant d < 20.5$, $20.5 \leqslant d < 30.5$, $30.5 \leqslant d < 50.5$

(b)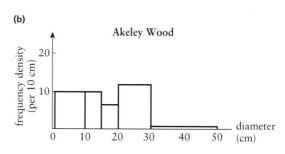

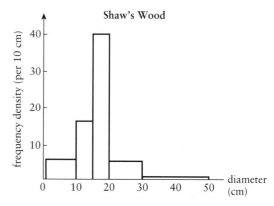

(c) Akeley Wood: 21–30; Shaw's Wood: 16–20
(d) For Akeley Wood there is a reasonably even spread of trees with diameters from 0.5 cm to 30.5 cm. For Shaw's Wood the distribution is centred about trees with diameter in the 16–20 cm interval. Neither wood has many trees with diameter greater than 30 cm.

2 (a)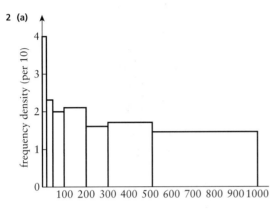

(b) The distribution has strong positive skew.

3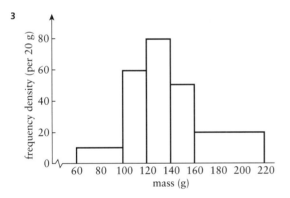

4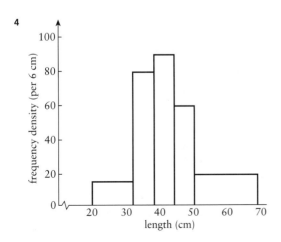

5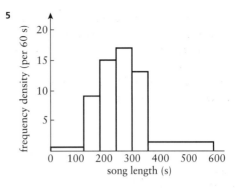

Exercise 2D (Page 34)

1 (a) (i) 6
  (ii) 4.5, 7.5
  (iii) 3
 (b) (i) 11
  (ii) 8, 14
  (iii) 6
 (c) (i) 26
  (ii) 23, 28
  (iii) 5
 (d) (i) 119.5
  (ii) 115.5, 126
  (iii) 10.5
 (e) (i) 5
  (ii) 2.5, 7.5
  (iii) 5
 (f) (i) 15
  (ii) 12.5, 17.5
  (iii) 5
 (g) (i) 275
  (ii) 272.5, 227.5
  (iii) 5
 (h) (i) 50
  (ii) 25, 75
  (iii) 50

2 (a) 74
 (b) 73, 76
 (c) 3
 (d)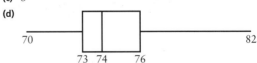

(e) On average the golfers played better in the second round; their average score (median) was four shots better. However, the wider spread of data (the IQR for the second round was twice that for the first) suggests some golfers played very much better but a few played less well.

3 (a)

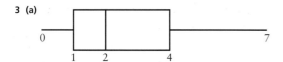

(b) The vertical line graph as it retains more data for this small sample.

4 (a)

$y \leqslant$	Cumulative frequency
49.5	1
59.5	6
69.5	13
79.5	17
89.5	19
99.5	20

(b)

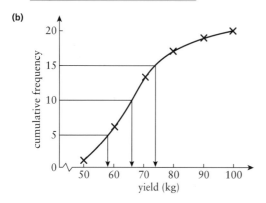

(c) 65 kg, 16 kg

(d)

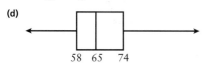

(e) 66 kg, 17.5 kg; the estimated values are quite close to these figures.

(f) Grouping allows one to get an overview of the distribution but in so doing you lose detail.

5 (a)

Status	Interval (pay £P/week)	Frequency (employees)	Interval length (×£10)	Frequency density (employees/£10)
Unskilled	$120 \leqslant P < 200$	40	8	5.0
Skilled	$200 \leqslant P < 280$	40	8	5.0
Staff	$280 \leqslant P < 400$	30	12	2.5
Management	$400 \leqslant P < 1000$	10	60	0.17

(b)

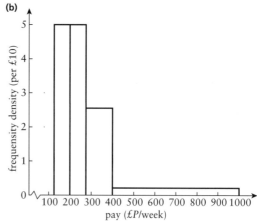

(c) Yes.

(d) The two displays emphasise different characteristics of the data. The pie chart draws attention to the relative proportion of workers in each category while the histogram emphasises the concentration of lower-paid workers.

6 (a)

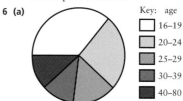

(b) Just turned 16 years of age and *almost* 20 years of age.

(c)

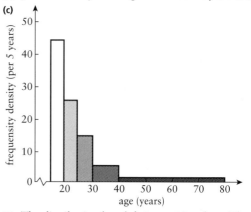

(d) The distribution has definite positive skew. The age range 16–19 is the most dangerous age range. The range 40–80 is the least dangerous. It is not possible to state which *single* age is the most/least dangerous, though it would be reasonable to estimate that the most dangerous, for example, was somewhere in the range 16–19.

(e) The histogram takes into account the different class widths, allowing the reader to see the *concentration* of ages.

Chapter 2

Exercise 2E (Page 37)

1 $Q_3 - Q_2 = £10.50$, $Q_2 - Q_1 = £13$, implying a slight negative skew.

2 $n = 87$

```
 1 | 8  represents 18 marks

 1 | 8 7
 2 | 6 9 2 6 5
 3 | 7 4 4 0 9 2 7 7 0 9 5 6
 4 | 4 5 3 9 0 4 6 4 9 5 0 8
 5 | 4 0 4 1 2 9 1 7 4 2 1 3 6 5
 6 | 6 6 9 8 8 0 1 6 2 6 9 7 5 4
 7 | 6 1 5 4 0 7 6 5 3 4 0
 8 | 0 7 2 7
 9 | 0 5 6 5 7 8 7 2 0 4 1
10 | 0 0
```

The distribution is symmetrical apart from a peak in the 90s. There is a large concentration of marks between 30 and 80.

3 (a) $n = 40$

```
 1 | 9  represents 19 years of age

 1 | 9 9 9 7
 2 | 8 4 0 8 6 2 5 6 2 6 6 9 3 3 1 8 1 3
 3 | 7 4 5 0 1 0
 4 | 0 5
 5 | 8 8 7
 6 | 5 6 9 5 5 7
 7 | 2
```

(b) The distribution is bimodal. This is possibly because those who hang-glide are the reasonably young and active (average age about 25 years) and those who are retired and have taken it up as a hobby (average age about 60).

4 (a) median = 16, IQR = $16.5 - 15 = 1.5$

(b)

(c) mean = 16.1

(d) The distribution is positively skewed – the boxplot shows the tail to the right.

5 (a) mode = 78

(b) lower quartile = 56, median = 70, upper quartile = 78

(c)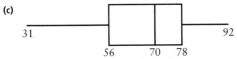

(d) Since the boxplot shows a tail to the left, the data are negatively skewed.

6 (a)

Length of leaf (mm)	Cumulative frequency	
	Strain A	Strain B
⩽ 9.5	3	1
⩽ 14.5	9	5
⩽ 19.5	20	11
⩽ 24.5	42	21
⩽ 29.5	77	37
⩽ 34.5	88	62
⩽ 39.5	94	82
⩽ 44.5	98	93
⩽ 49.5	99	97
> 49.5	100	100

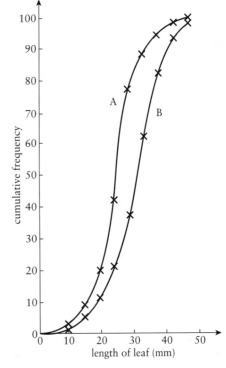

(b) (i) Strain B is better for longer leaves; median = 32.5 mm, compared with 25.1 mm for A.

(ii) Nothing much to choose between the strains for uniformity; the IQR for each is similar, for A: 8, for B: 9, so the level of variability is similar.

7 (a) Men in both categories (manual and non-manual) earn on average over £100 per week more than women.

(b) Manual (men): $IQR = £106$
Manual (women): $IQR = £59$
Earnings by men more variable, roughly twice as much.
Non-manual (men): $IQR = £183$
Non-manual (women): $IQR = £117$
Earnings by men, again, more variable.

8 89

9 (a) 7, 6, 4, 8
(b) 6.55, 5.7, 8.1
(c)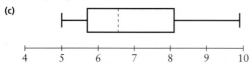
(d) The data are positively skewed.

10 (a) 49.6
(b)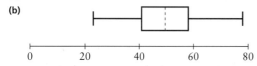
(c) The data are reasonably symmetrical.

Chapter 3

Exercise 3A (Page 45)

1 (a) mode = 45, mean = 39.6, median = 41
(b) bimodal, 116 and 132, mean = 122.5, median = 122
(c) mode = 6, mean = 5.3, median = 6

2 (a) (i) mode = 14 years 8 months, mean = 14 years 5.5 months, median = 14 years 6 months
(ii) Small data set so mode is inappropriate. You would expect all the students in one class to be uniformly spread between 14 years 0 months and 15 years, so any of the other measures would be acceptable.
(b) (i) mode = 4, mean = 3.85, median = 4
(ii) The mean does not give a very meaningful statistic. The mode is the most representative statistic though in such an example it should be no surprise that the median has the same value.
(c) (i) bimodal, 0 and 4, mean = 2.5, median = 3
(ii) The median probably gives the most representative data but the mean allows the researcher to retain the total quantity of beer drunk (mean × sample size) and this may be more useful depending on the aim of the survey.
(d) (i) mode = 0, mean = 52.8, median = 58
(ii) The median. Small sample makes the mode unreliable and the mean is influenced by outliers.
(e) (i) mode = 0 and 21 (bimodal), mean = 29.4, median = 20
(ii) Small sample so the mode is inappropriate; the mean is affected by outliers, so the median is the best choice.
(f) (i) no unique mode, mean = 3.45, median = 3.5
(ii) Anything but the mode will do. The distribution, uniform in theory, means that mean = median. This sample reflects that well.

Exercise 3B (Page 48)

1 (a) mode = 2
(b) median = 3
(c) mean = 3.24

2 (a) mode = 39
(b) median = 40
(c) mean = 40.3
(d) The sample has a slight positive skew. Any of the measures would do; however, the mean allows one to calculate the total number of matches.

3 (a) mode = 19
(b) median = 18
(c) mean = 17.9
(d) The outliers affect the mean. As the distribution is, apart from the extremes, reasonably symmetrical, the median or mode are acceptable. The median is the safest for a relatively small data set.

4 (a) mode = 1, median = 1, mean = 1.4
(b) the median

5 (a) mode = 1, mean = 2.1, median = 2
(b) the median

Exercise 3C (Page 56)

1 Mid-class values: 114.5, 124.5, 134.5, 144.5, 154.5, 164.5, 174.5, 184.5
Mean = 161.5 cm

2 (a) Mid-class values: 24.5, 34.5, 44.5, 54.5, 64.5, 79.5, 104.5
Mean = 48.5 minutes
(b) The second value seems significantly higher. To make the comparison valid the method of data

collection would have to be similar, as would the target children sampled.
3 (a) Mid-interval values: 4.5, 14.5, 24.5, 34.5, 44.5, 54.5, 64.5, 74.5, 84.5
Mean (stated age) = 29.7
(b) Add 0.5 to the mean age = 30.2 years
4 Mean = 43.1 cm
5 (a) Class mid-values: 25, 75, 125, 175, 250, 400, 750, 3000
Mean = 950.1 m
(b) The way in which these data are grouped seems to have a marked effect on the mean. This is probably because the distribution is so skewed.
6 (a) $59.5 \leqslant x < 99.5$
(b) 138.5 g

Exercise 3D (Page 64)

1 (a) Mean = 2.36
(b) Standard deviation = 1.49
2 Mean = 6.04, standard deviation = 1.48
3 (a) Steve: mean = 1.03, standard deviation = 1.05
Roy: mean = 1.03, standard deviation = 0.55
(b) On average they scored the same number of goals but Roy was more consistent.
4 Mean = 1.1, standard deviation = 1.24
5 Mean = 0.4, standard deviation = 0.4
6 (a) A: mean = 25 °C, standard deviation = 1.41 °C
B: mean = 25 °C, standard deviation = 2.19 °C
(b) Thermostat A is better. The smaller standard deviation shows it is more consistent in its settings.
(c) Mean = 24.8 °C, standard deviation = 1.05 °C
7 (a) Town route: mean time = 20 minutes, standard deviation = 4.60 minutes
Country route: mean time = 20 minutes, standard deviation = 1.41 minutes
(b) Both routes have the same mean time but the country route is less variable or more consistent.
8 (a) Yes. The value is more than two standard deviations above the mean rainfall.
(b) No. The value is less than one standard deviation below the mean rainfall.
(c) Overall mean rainfall = 1.62 cm, overall standard deviation = 0.135 cm
(d) 84.4 cm
9 (a) No. The harvest was less than two standard deviations above the expected value.
(b) The higher yield was probably the result of the underlying variability but that is likely to be connected to different weather patterns.

10 Mean = 6.5, standard deviation = 2.08 Combined data: mean = 6.76, standard deviation = 1.91
11 (a) Total weight = 3147.72 g
(b) $\sum x^2$ = 84 509.9868
(c) $n = 200$, $\sum x = 5164.84$, $\sum x^2 = 136\,549.913$
(d) Mean = 25.8 g, standard deviation = 3.98 g

Exercise 3E (Page 69)

1 Suggested code: $\dfrac{\text{mass} - 254.5}{4}$
Coded data: −3, −2, −1, 0, 1, 2
Mean = 252.34 g, standard deviation = 5.14 g
2 (a) $\bar{x} = -1.7$, $sd = 3.43$
(b) Mean = 94.83 mm, standard deviation = 0.343 mm
(c) −18 is more than four standard deviations below the mean value.
(d) New mean = 94.863 mm, new standard deviation = 0.255 mm
3 (a) £73.98, £20.09
(b) £86.93, £23.61
4 (a) Mean = 6.19, standard deviation = 0.484
(b) Mean = 6.68, standard deviation = 1.26
5 (a) No unique mode (5 and 6), mean = 5, median = 5
(b) 50 and 60, 50, 50
(c) 15 and 16, 15, 15
(d) 10 and 12, 10, 10
6 (a) $\bar{x} = -4.3$ cm, $sd = 13.98$ cm
(b) Mean length = 99.957 m, standard deviation in length = 0.140 m
(c) −47 is many more than two standard deviations from the mean.
(d) −2.05, 10.24

Exercise 3F (Page 71)

1 (a) Mean = £4200, median = £3700, mode = £3500
(b) Standard deviation = £2200
(c) £11 000 is more than two standard deviations above the mean and can therefore be regarded as an outlier.
(d) Mean in part (a) is per school and it is unlikely that each school will have the same number of pupils.
2 (a) $n = 14$

4	7 represents 47
4	7 9
5	9
6	8 2 7 6 8
7	3 0 4 2
8	4 0

Some negative skew, but otherwise a fairly normal shape.

(b) Mean = 67.07; standard deviation = 9.97
(c) $\bar{c} = \frac{5}{9}(\bar{f} - 32) = \frac{5}{9}(67.07 - 32) = 19.48$
$sd_c = \frac{5}{9} sd_f = \frac{5}{9} \times 9.97 = 5.54$

3 (a) Mean = 3.382, standard deviation = 2.327
(b) 4–6 in each case
(c) Standard deviation for A is greater than the standard deviation for B, since the data for A are more spread out.
(d) Player A is better; he or she knocks down most of the skittles at the first attempt.

4 (a) A, B and C appear to be compulsory subjects.
(b) Mean = 108.75, median = 116.5, mode = 207
Mode does not indicate central tendency; others do.
(c) Standard deviation = 70.49
(d) Total number of subject entries = 1740,
$\frac{1740}{207} = 8.4$ subjects per pupil
(e) $\frac{1740}{79} = 22.0$ pupils per class

5 (a) Mean = 1.2, standard deviation = 3.71
(b) 73.2, 3.71
(c) In the second round the golfers had a better average score and the scores were more consistent.
(d) Mean = 72.45, standard deviation = 3.41

6 (a) $n = 15$

3	40 represents £3.40
3	40 60 75 95
4	20 50 75
5	20 75
6	45 60
7	25
8	75
9	60
10	
11	
12	25

Median = £5.20, range = £8.85
(b) Mean = £6.00, standard deviation = £2.47
(c) A: mean = £6.30, standard deviation = £2.47
B: mean = £6.30, standard deviation = £2.59
(d) Both offers result in the same mean wage and therefore the same *total* wage bill (mean × n). Consequently the financial outlay in each case is the same so either scheme is acceptable to the management. The spread of wages, however, is greater for the 5% rise. Some workers would gain more by accepting this deal (those whose current wage was above the current mean wage). Of course, other workers would gain less by accepting this deal.

7 (a) Mode = 1
(b) The distribution is unimodal and has positive skew.
(c) Mean = 2, standard deviation = 1.70; the value 8 may be regarded as an outlier as it is more than two standard deviations above the mean.
(d) (i) Exclude it since a difference of 8 is impossible.
(ii) Check the validity and include if valid.
(e) Mean = 1.88, standard deviation = 1.48

8 (a) $5 \times 5 = 25$
(c) Mean = 69.55 mm, raw data are lost in a frequency distribution.
(d) 48 apples < 70 mm, 73 apples < 75 mm, so median is about 70.5.
(e) Distribution has negative skew and is unimodal.
(f) About 3 apples have a diameter less than mean − 2 × standard deviation = 51.55; about 7 apples have a diameter greater than mean + 2 × standard deviation = 87.55; 10 apples can be regarded as outliers.

9 (a) (i) 2.9, 6.0
(ii) 1.5, 3.8
(b) 2.5, 10.5, 30.5, 55, 72.5
Mean = 38.7

10 (a) Median = 30.6, $Q_1 = 25.9$, $Q_3 = 38.5$
(b) The distribution is positively skewed.
(c) With a mean around 30, some clients may take more than 60 minutes but none can take less than 0. Thus positive skewness is to be expected.
(d) Mean = 33.1, variance = 160.7
(e) The mean and standard deviation might be preferable, since they take all of the data into account.
(f)

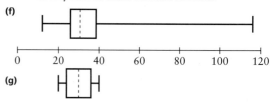

(h) The median and quartiles are similar for both data sets, implying that the average times and the location of the middle 50% of the data are much the same in each case.

The major difference is in the range, which is far greater in the first case. The first data set also exhibits positive skew, whereas the second is symmetrical.

Chapter 4

Exercise 4A (Page 85)

1. $\frac{66}{534}$, assuming each faulty torch has only one fault.
2. (a) $\frac{1}{6}$
   (b) $\frac{3}{6}$
   (c) $\frac{3}{6}$
   (d) $\frac{3}{6}$
3. (a) $\frac{12}{98}$
   (b) $\frac{53}{98}$
   (c) $\frac{45}{98}$
   (d) $\frac{42}{98}$
   (e) $\frac{56}{98}$
   (f) $\frac{5}{98}$
4. (a) $\frac{5}{2000}$
   (b) $\frac{1995}{2000}$
   (c) Lose £100, if all tickets are sold.
   (d) 25p
   (f) 2500
5. (a) 0.35
   (b) They might draw.
   (c) 0.45
   (d) 0.45
6. (a)

18	*E*	12		13	3	*O*	15	$\mathcal{E}$
	2		*S*			5		
10	4	1			19			
		16	9					
	6			7				
20	14	8		11		17		

   (b) (i) $\frac{10}{20}$
   (ii) $\frac{4}{20}$
   (iii) $\frac{10}{20}$
   (iv) $\frac{2}{20}$
   (v) $\frac{12}{20}$
   (vi) 0
   (vii) 1
   (viii) $P(E \cup S) = P(E) + P(S) - P(E \cap S)$
   (ix) $P(E \cup O) = P(E) + P(O) - P(E \cap O)$
7. (a) 0.22
   (b) 0.40
   (c) 0.10
   (d) 0.65
   (e) 0.35

Exercise 4B (Page 90)

1.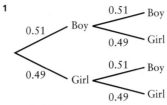
   (a) 0.2401
   (b) 0.5002
   (c) 0.4998

2.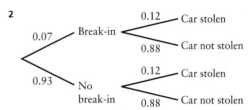
   (a) 0.0084
   (b) 0.1732
   (c) 0.1816

3. $\frac{1}{12}$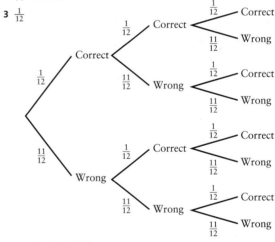
   (a) (i) 0.00058
   (ii) 0.77
   (iii) 0.020
   (b) (i) 0.0052
   (ii) 0.52
   (iii) 0.094

4. 0.93

5.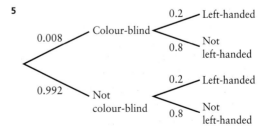

Statistics 1 — Answers

(a) 0.0016
(b) 0.0064
(c) 0.2064
(d) 0.7936

6 $\frac{3}{4}$

7 (a) 0.2436
  (b) 0.7564
  (c) 0.2308
  (d) 0.4308

8 (a) 0.0741
  (b) 0.5787
  (c) 0.5556

9 For a sequence of events you multiply the probabilities. However, $\frac{1}{6} \times \frac{1}{6} \times \frac{1}{6} \times \frac{1}{6} \times \frac{1}{6} \times \frac{1}{6}$ gives the probability of six 6s in six throws. To find the probability of at least one 6 you need $1 - P(\text{no 6s})$ and that is $1 - \frac{5}{6} \times \frac{5}{6} \times \frac{5}{6} \times \frac{5}{6} \times \frac{5}{6} \times \frac{5}{6} = 0.665$.

10 0.5833

11 (a)

Green die

+	1	2	3	4	5	6
1	2	3	4	5	6	7
2	3	4	5	6	7	8
Red 3	4	5	6	7	8	9
die 4	5	6	7	8	9	10
5	6	7	8	9	10	11
6	7	8	9	10	11	12

  (b) $\frac{3}{36}$
  (c) 7
  (d) The different outcomes are not all equally probable.

12 (a) $\frac{1}{2}$
   (b) $\frac{1}{8}$
   (c) $\frac{3}{8}$
   (d) 62.5 pence

13 0.31

EXERCISE 4C (Page 96)

1 (a) 0.6
  (b) 0.556
  (c) 0.625
  (d) 0.047
  (e) 0.321
  (f) 0.075
  (g) 0.028
  (h) 0.0022
  (i) 0.000 95
  (j) 0.48

2 (a) (i) 0.160

   (ii) 0.120
   (iii) 0.400
   (iv) 0.160
   (v) 0.003
   (vi) 0.088
   (vii) 0.018

  (b) Those sentenced for motoring offences would probably have shorter sentences than others so are likely to represent less than 2% of the prison population at any time.

3 (a) $\frac{35}{100}$
  (b) $\frac{42}{100}$
  (c) $\frac{15}{65}$

4 (a) $\frac{1}{6}$
  (b) $\frac{5}{12}$
  (c) $\frac{2}{5}$

5 (a) $\frac{7}{15}$
  (b) $\frac{11}{21}$
  (c) $\frac{5}{21}$
  (d) $\frac{5}{11}$

6 (a) 30
  (b) $\frac{7}{40}$
  (c) $\frac{7}{10}$
  (d) $\frac{7}{15}$

7 (a) $\frac{7}{10}$
  (b) $\frac{19}{40}$
  (c) $\frac{10}{19}$

8 (a) 0.479
  (b) 0.477
  (c) 0.222
  (d) 0.461

9 (a) 0.5 and 0.875
  (b) $P(B \mid A) \neq P(B)$ and $P(A \mid B) \neq P(A)$ so the events A and B are not independent.

10 (a)

	Hunter dies	Hunter lives	
Quark dies	$\frac{1}{12}$	$\frac{5}{12}$	$\frac{1}{2}$
Quark lives	$\frac{2}{12}$	$\frac{1}{3}$	$\frac{1}{2}$
	$\frac{1}{4}$	$\frac{3}{4}$	1

   (b) $\frac{1}{12}$
   (c) $\frac{5}{12}$
   (d) $\frac{5}{6}$

EXERCISE 4D (Page 99)

1 (a) 0.4
  (b) 0.4
  (c) 0.2
  (d) $0.5 \times 0.3 = 0.15$, not 0.2, so they are not independent.

190

2 (a)

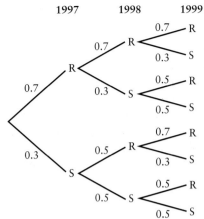

(b) 0.372
(c) 0.395
(d) 8

3 (a)

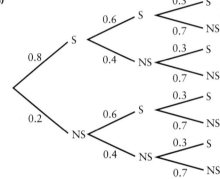

(b) 0.056
(c) 0.332
(d) 0.675
(e) $\frac{14}{41}$

4 (a)

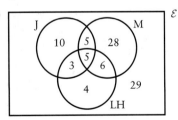

Key:
J = Juniors
M = Males
LH = Left-handed players

(b) (i) $\frac{1}{4}$
  (ii) $\frac{1}{6}$
  (iii) $\frac{28}{45}$
  (iv) $\frac{4}{5}$
  (v) $\frac{19}{24}$
  (vi) $\frac{10}{39}$

5 (a) $\frac{11}{15}$
(b) $\frac{3}{8}$

6 (b) $\frac{29}{72}$
(c) $\frac{18}{43}$

7 (a)

	Fruit tree	Other tree	Total
Bird's nest	2	4	6
No nest	5	9	14
Total	7	13	20

(b) $\frac{9}{10}$
(c) $\frac{2}{7}$

8 (a) $\frac{11}{24}$
(b) $\frac{11}{60}$
(c) $\frac{43}{120}$
(d) $\frac{49}{144}$

9 (a) 0.28
(b) 0.3
(c)

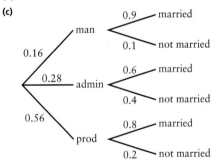

(d) 0.76
(e) $0.448 \div 0.76 = 0.589$

10 (a) $\frac{1}{4}$
(b) $\frac{5}{14}$
(c) Yes, they are independent.
(d) $\frac{1}{7}$, $P(A|C) \neq P(A)$
(e) $\frac{7}{18}$

Chapter 5

Exercise 5A (Page 108)

1 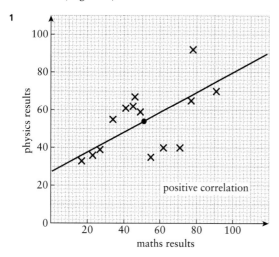 positive correlation

2 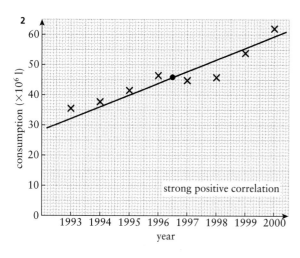 strong positive correlation

3 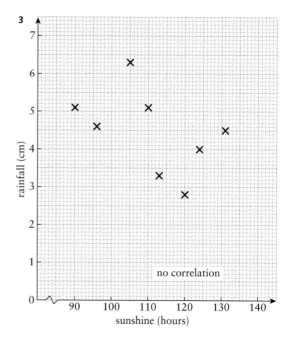 no correlation

4 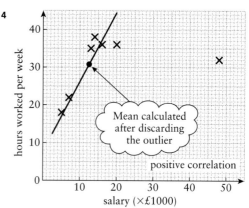 Mean calculated after discarding the outlier — positive correlation

5 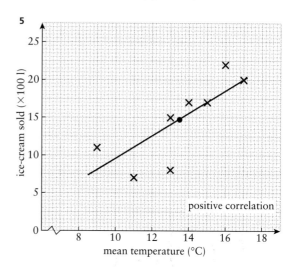 positive correlation

6

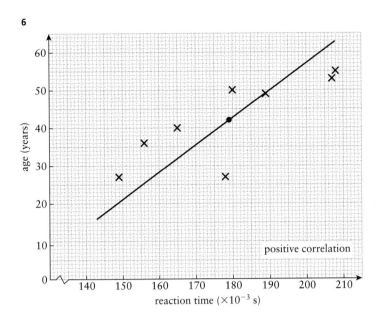

positive correlation

Exercise 5B (Page 114)

1. (a) −0.8
   (b) 0
   (c) 0.8
2. −0.96
3. 0.704
4. −0.924
5. −0.635
6. −0.128
7. (a) 0.715
   (b) It seems that performance in high jump and long jump have positive correlation.
8. 0.850
9. 0.946
10. (a) 0.491
    (b)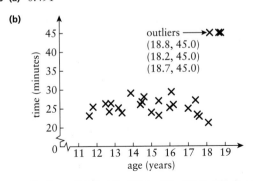

    Outliers: (18.8, 45), (18.2, 45), (18.7, 45), it seems as though these girls stopped for a rest.

Exercise 5C (Page 122)

1. $19x + 50y = 1615$, 27.7
2. $1.446x + y = 114.38$, 53.7
3. (a)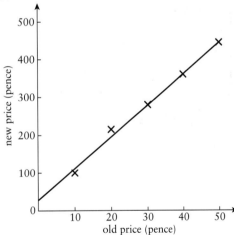
   (b) $y = 8.45x + 27.5$
   (c) Value of this stamp not compatible with earlier part of question.

4 (a)

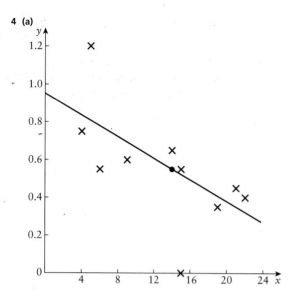

(b) $0.029x + y = 0.925$
(c) 0.432
(d) Stem density is outside domain of validity. If $x > 32$ model predicts that y is negative.

5

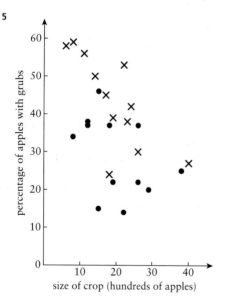

$1.013x + y = 64.25$, 44%. Second treatment seems to produce fewer grubs but linear correlation is not so strong as before.

6 (a)

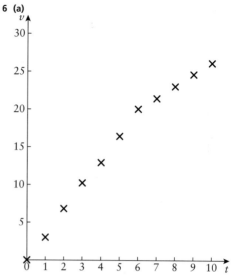

(b) $v = 2.68t + 1.54$
(c) 0.988
(d) Modelling data by a single straight line assumes that the correlation is linear. However, looking at the scatter diagram there is a possibility that r is proportional to a power of t thus making the correlation non-linear.

7 (a)

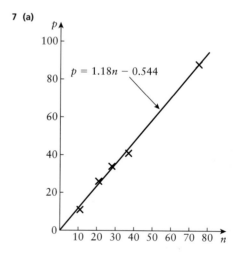

(b) $p = 1.18n - 0.544$; £17
(c) 0.998; Both correlation coefficient and scatter diagram suggest almost perfect positive correlation.

8 (a) -0.926
(b) $0.752x + y = 74.97$

9 (a)

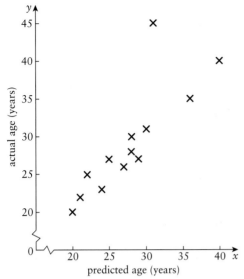

(b) $y = 1.031x + 0.528$

(c) Predictions seem very accurate despite the data for person G.

10 (a)

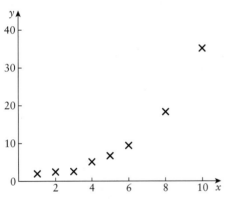

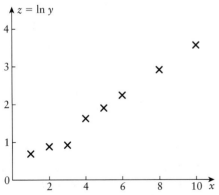

(b) The line of regression of y on x is not very accurate because the correlation is not linear. The line of regression of z on x is much more accurate.

(c) 12.8

Exercise 5D (Page 128)

1 (a) 0.3377

(b) Although the point is unusual it should not be excluded. It is a valid member of the data set. This point does reduce the value of the pmcc.

2 (a)

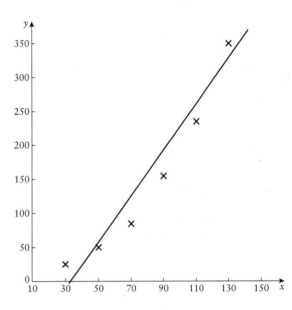

(b) $y = -107.1 + 3.21x$

(c) 214.3, 375

214.3 is likely to be accurate as it is obtained by interpolation.

375 is likely to be inaccurate as it is obtained by extrapolation.

(d) From the graph it seems that a curve would offer a better fit, perhaps a quadratic.

3 (a)

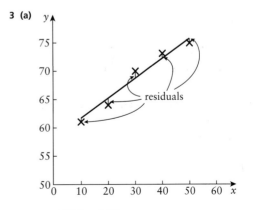

(b) $y = 57.5 + 0.37x$

(c) 70.5 g
Estimates should be reliable from 10–50 °C, i.e. over the range of the data.
(d) −0.2, −0.9, 1.4, 0.7, −1
(e) The regression line minimises the sum of the squares of the residuals.

4 (a) 0.4735
(b) Elliptical distribution of points
(c) It is a tendency or trend only. There are specific instances where a taller student is lighter than another.

5 (a)

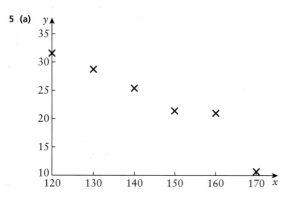

A straight line looks a reasonable model, especially in the earlier part of the range.
(b) $y = 77.3 − 0.374x$
(c) 23.1, 10.0
The first of these is more likely to be accurate, as it is interpolated; the second is extrapolated beyond the range of the data.

6 (a) 0.9564
Looks like good evidence of linear correlation, since points are elliptical.
(b) $y = −18.3 + 15.98x$
(c) 48.8
This should be quite reliable because it lies within the range of the data (interpolation).
(d) The equation should *not* be used when $x = 15$ since this would involve extrapolation well beyond the range of the data.

7 (a)

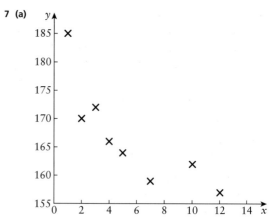

(b) q indicates the growth per year (positive) or decline (negative) in the fish caught.
(c) The model implies that no fish at all will be caught in about 94 years, i.e. the stock is declining.
(d) $y = 157.8 + 27.96x$
(e) As $t \to \infty$ $x \to 0$ and so y will settle to a long term limit of 157.8.
(f) a is the number of fish caught once the population has stabilised.

8 (a), (d)

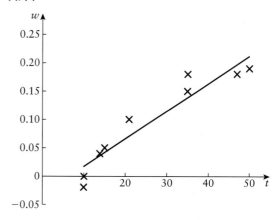

(b) $S_{tt} = 1980$, $S_{tw} = 9.697$
(c) $w = −0.031 + 0.0049t$
(e) a indicates the expected weight gain for a rat with no fat; b indicates the rate of weight gain per unit of fat.
(f) The model looks reasonable because of the approximately elliptical distribution of the points, but perhaps a parabola or other curve might fit better.

Chapter 5

(g) $w = 0.851$

This is not a reliable prediction since it requires extrapolation well beyond the range of the data.

(h) For rats that do not shiver at all, t is indefinitely large, implying an indefinitely large weight gain, which is clearly unrealistic.

9 (a) -0.8959

(b) The population has gone up in some years, e.g. 1960–62 and again in 1980–82.

10 (a) $y = 14.55 + 1.0227t$

(b) a indicates the temperature in the cool room.

(c) 19.15, 35.0

(d) 19.15 is likely to be the more reliable of these, since it is interpolated within the range of data.

(e)
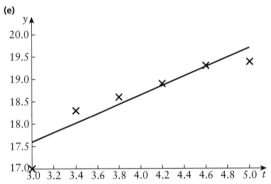

(f) If the linear model were used for a long period of time it would imply an unreasonably high temperature for the warm room.

(g) A curve would be better, as (3.0, 17.0) could then fit more closely; also y should tend to a constant value (e.g. 19.5) for large values of t.

Chapter 6

Exercise 6A (Page 139)

1 $k = \frac{1}{10}$

2 Number of turns needed to obtain a '6' on a die.

3 (a) $a = 0.4$

(b) 0.3

4

X	0	1	2	3	4	5
Probability	$\frac{1}{32}$	$\frac{5}{32}$	$\frac{10}{32}$	$\frac{10}{32}$	$\frac{5}{32}$	$\frac{1}{32}$

5 $c = \frac{1}{28}; \frac{3}{28}$

6

X	2	3	4	5	6	7	8	9	10	11	12
Probability	$\frac{1}{36}$	$\frac{2}{36}$	$\frac{3}{36}$	$\frac{4}{36}$	$\frac{5}{36}$	$\frac{6}{36}$	$\frac{5}{36}$	$\frac{4}{36}$	$\frac{3}{36}$	$\frac{2}{36}$	$\frac{1}{36}$

7 (a)

Y	0	1	2	3	4	5
Probability	$\frac{6}{36}$	$\frac{10}{36}$	$\frac{8}{36}$	$\frac{6}{36}$	$\frac{4}{36}$	$\frac{2}{36}$

(b) $\frac{2}{3}$

8 (a)

Z	0	1	2	3	4
Probability	$\frac{1}{16}$	$\frac{4}{16}$	$\frac{6}{16}$	$\frac{4}{16}$	$\frac{1}{16}$

(b) $\frac{5}{16}$

9

X	0	1	2	3
Probability	0.1667	0.5	0.3	0.0333

10

Number of men	0	1	2	3
Probability	0.122	0.441	0.367	0.070

11 (a)

X	1	2	3	4	6	8	9	12	16
Probability	$\frac{1}{16}$	$\frac{2}{16}$	$\frac{2}{16}$	$\frac{3}{16}$	$\frac{2}{16}$	$\frac{2}{16}$	$\frac{1}{16}$	$\frac{2}{16}$	$\frac{1}{16}$

(b) 0.25

12

Number of red cards	0	1	2	3	4
Probability	0.055	0.25	0.39	0.25	0.055

13 (a) $k = 0.08$

X	0	1	2	3	4
Probability	0.2	0.24	0.32	0.24	0

(b)

Number of chicks surviving	0	1	2	3
Probability	0.35104	0.44928	0.18432	0.01536

14 (a) $a = 0.42$

(b) $k = \frac{1}{35}$

(c) Algebraic model is not particularly accurate. No.

15 (a) $\frac{1}{1296}$

(b) $\frac{1}{81}$

(c) $\frac{65}{1296}$

(d) $\frac{671}{1296}$; 6 is the most likely value because $P(X = 6) > \frac{1}{2}$

Exercise 6B (Page 146)

1 1.5

2 2.7

3 $P(X = 4) = 0.8, P(X = 5) = 0.2$

4

Y	50	100
Probability	0.4	0.6

5 3.22

Statistics 1 — Answers

6 (a) 0.65
  (b) 0.5
  (c) 0.4
7 (a) loss of 2.5p
  (b) loss of 7.5p
  (c) loss of £2.50
8 (a) £168
  (b) £2016
9 (a) (i) 2.79
     (ii) 8.97
     (iii) 20.94
10 (a) $w = 21$
   (b) $w = 27.67$

11 (a)
X	2	3	4	5	6	7	8	9	10	11	12
$P(X)$	$\frac{1}{36}$	$\frac{2}{36}$	$\frac{3}{36}$	$\frac{4}{36}$	$\frac{5}{36}$	$\frac{6}{36}$	$\frac{5}{36}$	$\frac{4}{36}$	$\frac{3}{36}$	$\frac{2}{36}$	$\frac{1}{36}$

(b)
Y	2	4	6	8	10	12
$P(Y)$	$\frac{1}{6}$	$\frac{1}{6}$	$\frac{1}{6}$	$\frac{1}{6}$	$\frac{1}{6}$	$\frac{1}{6}$

(c) (i) $E(X) = E(Y) = 7$
    (ii) Range of X = range of Y = 10
    (iii) Mode of $X = 7$, Y has no mode since all the outcomes are equally likely.

12 (a) 3 coins
   (b) £1.30

13
p_1	p_2	p_3	p_4	p_5	p_6
$\frac{1}{36}$	$\frac{3}{36}$	$\frac{5}{36}$	$\frac{7}{36}$	$\frac{9}{36}$	$\frac{11}{36}$

(a) 0.0405
(b) 0.1118

14 $\frac{4}{35}, \frac{18}{35}, \frac{12}{35}, \frac{1}{35}, \frac{£(54+r)}{35}, r = 50$

Exercise 6C (Page 151)

1 (a) (i) $E(X) = 3.1$
     (ii) $Var(X) = 1.29$
2 (a) (i) $E(X) = 0.7$
     (ii) $Var(X) = 0.61$
4 (a) $E(2X) = 6$
  (b) $Var(3X) = 6.75$
5 (a) $k = 0.1$
  (b) 1.25 eggs
  (c) 0.942
6 (a) $E(Y) = 1.2857$
  (b) $Var(Y) = 0.49$
7 (a) 10.9, 3.09
  (b) 18.4, 111.24

8 (a) 2
  (b) 1
  (c) 9
9 £5.47, 0.089

10
S	1	2	3	6	10
Probability	$\frac{1}{6}$	$\frac{1}{3}$	$\frac{1}{6}$	$\frac{1}{6}$	$\frac{1}{6}$

4, $9\frac{2}{3}$

11 (a) 2.3, 0.41
(b)
$S_1 + S_2$	2	3	4
Probability	0.51	0.45	0.04

2.53

12 (b)
X	0	1	2	3	4	6
Probability	$\frac{1}{4}$	$\frac{1}{3}$	$\frac{1}{9}$	$\frac{1}{6}$	$\frac{1}{9}$	$\frac{1}{36}$

(c) $1\frac{2}{3}, 2\frac{5}{18}$

13
X	6	7	8	9	10
Probability	$\frac{6}{72}$	$\frac{24}{72}$	$\frac{24}{72}$	$\frac{16}{72}$	$\frac{2}{72}$

0.975, 0.6397

14 2.449, 2.574

15
$\overline{N}$	2	3	4	5	6	7	8
Probability	$\frac{1}{16}$	$\frac{2}{16}$	$\frac{3}{16}$	$\frac{4}{16}$	$\frac{3}{16}$	$\frac{2}{16}$	$\frac{1}{16}$

5, 2.5

16 (a)

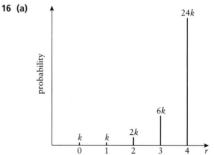

(b) 3.5, 0.0897
(c) 0.5346
(d) 0.932

17 (a)
X	0	1	2	3	4
Probability	0.003	0.1	0.133	0.167	0.167

X	5	6	7	8
Probability	0.167	0.122	0.089	0.022

(b) 3.9, 3.93
(c) $k = \frac{1}{84}$
(d) 4, 3
(e) Yes

Chapter 6

18 (a)

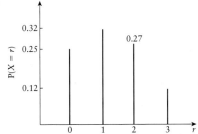

(b) 1.34, 1.044
(c) (i) 0.0625
   (ii) 0.1753. Assume the number of goals scored is independent of previous results.
(d) Injuries to players, new signings, promotion to a new league, etc. Model does not allow more than 3 goals to be scored.

Exercise 6D (Page 157)

1 (a)

r	0	1	2	3	4
P(X = r)	$\frac{1}{5}$	$\frac{1}{5}$	$\frac{1}{5}$	$\frac{1}{5}$	$\frac{1}{5}$

(b) 2, 2
(c) 5, 8

2 (a) Discrete uniform distribution

r	1	2	3	4
P(X = r)	$\frac{1}{4}$	$\frac{1}{4}$	$\frac{1}{4}$	$\frac{1}{4}$

(b) 2.5, 1.25
(c)

	1	2	3	4
1	2	3	4	5
2	3	4	5	6
3	4	5	6	7
4	5	6	7	8

e.g. $P(Y = 2) = \frac{1}{16}$ but $P(Y = 4) = \frac{3}{16}$

3 (a)

r	0	1	2	3	4	5	6	7	8	9
P(X = r)	0.1	0.1	0.1	0.1	0.1	0.1	0.1	0.1	0.1	0.1

(b)

r	0	1	2	3	4	5	6	7	8	9
P(X ⩽ r)	0.1	0.2	0.3	0.4	0.5	0.6	0.7	0.8	0.9	1.0

(c) 4.5, 8.25
(d) 0.4 × 400 = 160

4 (a) p = 0.25
(b) 1.5, 1.25
(c) The model assigns equal probabilities to X = 0, 1, 2, 3 and does not allow X to be greater than 3.

5 (a)

r	3	5	7	9
P(X = r)	0.25	0.25	0.25	0.25

(b) 6, 5
(c) 11, 1.25

Exercise 6E (Page 158)

1 (a) $P(Y = r) = \frac{1}{6}, r = 1, 2, 3, 4, 5, 6$
(b) Y has a rectangular (uniform) distribution.
(c) 23
(d) 46.67

2 (a) P(X = 3) is less than P(X = 2) for example, since the latter may be achieved in many different ways, so the distribution cannot be uniform.
(b) $P(X = 3) = (\frac{1}{6})^3 = 0.0046$
(c) $P(X = 2) = 0.0425, P(X = 3) = 0.0046$
(d) 0.473, 0.3619

3 (a) $\frac{1}{2}$
(b) $\frac{17}{36}$
(c) $\frac{5}{6}$

4 (a) $\alpha = 0.3, \beta = 0.2$
(b) 0.6
(c) 2.36
(d) −2.6
(e) 9.44

5 (a)

r	0	1	2	3
P(X ⩽ r)	0.1	0.5	0.8	1.0

(b) If L ⩽ 2 then each separate shot is also ⩽ 2.
∴ $P(L \leqslant 2) = [P(X \leqslant 2)]^3$
0.125; 0.001, 1

(c)

r	0	1	2	3
P(L = r)	0.001	0.124	0.387	0.488

(d) 2.362, 0.485

6 (a) 0.3, 0.147, 0.343
(b)

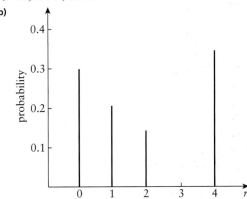

(c) 1.876, 1.66
(d) 0.363

7 (a) 0.4
(b) 0.8
(c) 2.6
(d) 1.44
(e) 15.6

8 (a) $P(X = 2) = \frac{1}{2} \times \frac{1}{6} + \frac{1}{3} \times \frac{1}{3} + \frac{1}{6} \times \frac{1}{2} = \frac{5}{18}$

(b)

r	0	1	2	3	4
$P(X = r)$	$\frac{1}{4}$	$\frac{1}{3}$	$\frac{5}{18}$	$\frac{1}{9}$	$\frac{1}{36}$

(c) $1\frac{1}{3}$
(d) $2\frac{8}{9} - 1\frac{7}{9} = \frac{10}{9}$
(e) $E(Z) = E(X) - E(Y) = 0$
$Var(Z) = Var(X) + Var(Y) = \frac{20}{9}$

9 (a) $P(Q = 2) = \frac{1}{4} \times \frac{2}{3} + \frac{1}{2} \times \frac{1}{3} = \frac{1}{3}$

(b)

r	1	2	3	4	6
$P(Q = r)$	$\frac{1}{12}$	$\frac{1}{3}$	$\frac{1}{12}$	$\frac{1}{3}$	$\frac{1}{6}$

(c) $E(Q) = 1 \times \frac{1}{12} + \ldots + 6 \times \frac{1}{6} = \frac{10}{3}$
(d) $\frac{43}{18} = 2.389$

10 (a)

r	−5	−3	−1	2	4	6
$P(X = r)$	$\frac{1}{6}$	$\frac{1}{6}$	$\frac{1}{6}$	$\frac{1}{6}$	$\frac{1}{6}$	$\frac{1}{6}$

(b) 0.5
(c) $Var(X) = 15\frac{1}{6} - 0.5^2 = \frac{179}{12}$
(d) $E(X) > 0$ implies that this is a growth model.
(e) $\frac{1}{1296}$

Chapter 7

Exercise 7A (Page 170)

1 (a) 0.8413
(b) 0.0228
(c) 0.1359
2 (a) 0.0668
(b) 0.6915
(c) 0.2417
3 (a) 0.0668
(b) 0.1587
(c) 0.7745
4 (a) 0.0062
(b) 0.5
(c) 0.5468
5 0.4435
6 (a) 0.1587
(b) 0.2417
(c) 0.5539
7 (a) B
(b) A
8 13.6%, 0.0185

Exercise 7B (Page 174)

1 (a) 0.0668
(b) 0.8664
(c) 0.6853
(d) 1.314 m
2 0.0401
(a) 0.4593
(b) 0.003
3 (a) 5.48%
(b) (i) 24 900 km
(ii) 2041 km
4 (a) 78.65%
(b) 5.255, 0.053
5 20.05, 0.02215, 82%, 18%
6 (a) 14.87
(b) 0.0668
(c) 1.56
7 (a) 16%
(b) 82.4 g
8 (a) 0.1465
(b) 0.4400
(c) 0.3345
(d) 8.27 and 40 s
9 (a) (i) 0.692
(ii) 0.308
(b) 0.446
(c) 32 days
10 (a) 18.7
(b) 9.6
(c) They are very different, implying that the marks are not normally distributed.

Exercise 7C (Page 176)

1 (a) 0.5987
(b) 0.4649
(c) 84.13%
2 (a) On average I pay early.
(b) 8%
(c) 0.3343

Chapter 7

3 (a) 89%
  (b) 0.5981
  (c) Not valid as intervals are not independent.
4 (a) 16%
  (b) 81.7°
5 (a) 0.1056
  (b) $z = 0.7$ so $k = 11.2$
6 (a) $73.8 - \mu = 1.9600\sigma$, $51.335 - \mu = -0.2533\sigma$
  (b) 53.9, 10.15
7 (a) 0.5309
  (b) 4340
  (c) Only 0.6% of children score above 10, so in practice it represents the maximum score for the vast majority.
  (d) 113 or more.
8 (a)

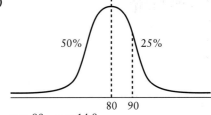

$\mu = 80 \quad \sigma = 14.8$
  (b) 0.004
  (c) $X \leqslant 60$, which is less than the mean
  (d) $\mu + 2\sigma = 110 \Rightarrow 3.10\,\text{pm}$
  (or $\mu + 3\sigma = 125 \Rightarrow 2.55\,\text{pm}$)
9 (a) 0.0668
  (b) 0.7745
  (c) 0.819
  (d) $\mu > 266.45$ so say 267 ml
10 (b) 0.8664
  (c) $(0.1336)^2 = 0.0178$

Index

average 42

bar chart 14–15, 41
bimodal distribution 44
bivariate data 104
   dangers of extrapolation 117–118
   line of best fit 104, 108
   product moment correlation 109–112
box and whisker plot 29–33, 41

categorical data 5, 41
central tendency 27–28, 42, 76
class boundary 21, 24, 55
class interval 56
class width 21–22
coding 67–69, 76, 126–127
complement 79–80
compound bar chart 15–16
conditional probability 93–96
continuous data 6, 20, 54–56
controlled variable 105
correlation 104–118
   interpreting 106–107, 117–118
   linear correlation 104
   negative correlation 106
   non-linear correlation 117
   positive correlation 106
correlation coefficient 109–112
cumulative distribution function 137, 162
cumulative frequency 30–33

data
   categorical or qualitative 5, 41
   continuous 6, 41, 54–56
   discrete 6, 41, 51
   grouped 23, 33, 50
   numerical or quantitative 5
   raw 7
dependent 89, 94, 105
discrete data 6, 41, 51
dispersion, measures of 58–64
distribution
   bimodal 9
   frequency 46–47
   probability 136–137

   shape of 9
   skewed 3, 10
   uniform 2–4, 156–157, 162
   unimodal 9

events
   complement of 79–80
   exhaustive 136
   mutually exclusive 83–84, 103
expectation 80, 142–146
experimental probability 1–2
explanatory (independent) variable 105, 118

frequency 20
   cumulative 30–33
   density 20–25, 41
   distribution 46–47
   table 42, 47

grouped data 23, 22, 50

histogram 9, 20–25, 41

independent 89, 93, 103, 105
interquartile range 29, 41, 59
intersection 84

least squares regression line 118–120, 134
linear correlation *see* correlation
line of best fit 104

mean 42–43, 67–69, 76
   estimating 52–54
median 27, 41, 43, 76
modal class 9, 76
mode 9, 44, 76
model 1–4
mutually exclusive 83–84, 103

non-linear correlation *see* correlation
normal distribution 3–4, 163–179
   inverse normal tables 172–173
   normal tables 164–165
   standardising 165, 179
   z-value 164

outlier 8, 29–30, 41, 63–64

Pearson, Karl 112
percentiles 32–33
pie chart 17, 41
probability 6, 77–103
   conditional 93–96
   distribution 136–137
   experimental 1–2
   of certainty or impossibility 1, 78–79
   theoretical 1–2
product moment correlation 109–112, 134

qualitative data 5, 41
quantitative data 5
quartile 27, 41
   spread 29

random variable 6, 135
range 59–60, 76
regression line *see* least squares regression line
residuals 119
response (dependent) variable 105, 118
sample space diagram 89–90
scatter diagrams 104, 106–107
sigma notation 42
skewness 10, 24
spread, measures of 58–64
standard deviation 60–64, 67–69, 76
stem and leaf diagram 10–12, 23, 41
stemplot 10

tally 7–8
theoretical probability 1–2
tree diagram 87–88
trial 1, 77

uniform distribution 2–2, 156–157, 162
union 81–82

variance 60–64, 67–69, 76
vertical line graph 14–15, 23, 41